Mental Math

Tricks To Become A Human Calculator

By Abhishek V.R

Ofpad – The School of Genius

Copyright

Copyright © Abhishek V.R Ofpad – The School of Genius

All rights reserved. No part of this book may be reproduced in any form or by any electronic or mechanical means, including information storage and retrieval systems, without written permission from the author, except in the case of a reviewer, who may quote brief passages embodied in critical articles or a review. Trademarked names appear throughout this book. Rather than use a trademark symbol with every occurrence of a trademarked name, names are utilised editorially, with no intention of infringement of the respective owner's trademark. The information in this book is distributed on an "as is" basis, without warranty. Although every precaution has been taken in the preparation of this work, neither the author nor the publisher shall have any liability to any person or entity concerning any loss or damage caused or alleged to be caused directly or indirectly by the information contained in this book.

Dedication

All that I am & all that I will ever be, I owe it to my mother.

This book is also dedicated to all members of Ofpad.com, who push the boundaries of their intelligence every day.

Contents

Foreword 1
Chapter 1 - What Is Ofpad Math System? 5
Chapter 2 - DS Method of Checking 9
    Rules To Calculate Digit Sum 9
    Rules To Apply DS Method of Checking 11
    Checking Addition 12
    Checking Subtraction 12
    Checking Multiplication 15
    Checking Division 16
    Tip – Reduce Running Total to Single Digits 20
    Limitations of the DS Method 21
    Summary 21
Chapter 3 - DD Method of Checking 23
    Rule To Apply DD Method of Checking 23
    Odd & Even Digits in Digit Difference 24
    Checking Addition 25
    Checking Subtraction 27
    Checking Multiplication 30
    Checking Division 33
    DS Method VS DD Method 36
Chapter 4 - Mental Math Q&A Community 39
Chapter 5 - Introduction to Mental Math 45
    Course Structure 45

    Defining a Multiplicand & a Multiplier 46

    Multiplication By 11 47

    Carrying Over In Mental Math 51

    Exercises 59

    Answers 59

Chapter 6 - The Inefficient Way to Do Math 61

Chapter 7 - Introducing The LR Method 65

Chapter 8 - Addition & Subtraction 69

    LR Addition 69

    Rounding Up Before Calculating 74

    LR Addition With Rounding Up 78

    When Should You Round Up For Addition? 84

    Addition Exercises 86

    Addition Answers 86

    LR Subtraction 87

    LR Subtraction With Rounding Up 92

    When Should You Round Up For Subtraction? 97

    Subtraction Exercises 99

    Subtraction Answers 100

Chapter 9 – Remembering Numbers 101

Chapter 10 - LR Multiplication 103

    One-digit Multiplier 104

    One-digit Multiplier With Rounding Up 110

    LR Multiplication – Two Digit Multiplier 117

    LR Multiplication – Two Digit Multiplier With Round Up 123

    LR Multiplication After Factoring 130

Chapter 11 - Stem Method... 139

    Stem Multiplication 139

    One-by-One Multiplication... 140

    Learn Multiplication Tables? 148

    Two-by-Two Multiplication 149

    Choosing The Stem 158

    Distance From The Stem 158

    When to Use Stem Method?.. 167

Chapter 12 - Math Anxiety 169

Chapter 13 - Squaring 171

    Squaring Numbers Ending With 5 171

    Squaring Any Number 174

    Exercises 180

    Answers 181

Chapter 14 - The Bridge Method 183

Chapter 15 - Vitruvian Man Method 195

Chapter 16 - UT Method 209

    UT Method Concepts 209

    Practice Exercise 209

    Pair Products 211

    Applying UT Method 218

Chapter 17 - LR Division 235

Definition 235
Division Refresher 236
LR Division 243
LR Division - Factoring 243
Divisor Ends With 5 252
Factoring Fails with Prime Numbers 259
LR Division With Rounding 261
Rounding With Reminder 269
Combining Methods 272
Chapter 18 - FP Division 277
Flag & Pole 277
Applying Flag Pole Method 278
Chapter 19 - Eliminating Skill Atrophy 295
Chapter 20 - Conclusion 297
About the Author 299

Foreword

I want to begin by telling you a story. This is a story that I am embarrassed to share. Yet it is one you need to hear because it will reveal how you too can do math in your head.

I wasn't always good with numbers, and I certainly wasn't smart growing up. In fact, I HATED math.

One beautiful morning, I was in the backseat of my cousin's motorcycle enjoying the cool breeze hitting my face.

My summer was coming to an end, and my cousin was dropping me at the railway station. I was returning home after enjoying the monsoon rains. Life was good until it decided to punch in the face.

This was a critical day of my life. In another 30 minutes, the results of my exam were going to be out. This was important because how much I scored in this exam determined whether or not I went to college.

In school, I wasn't very serious about my grades. My entire time was either spent playing video games or talking to my high school girlfriend on the phone.

I had no real goal or purpose. If you had asked me what I wanted to become, I would have told you, "I am still figuring that out". I was just an average guy going with the flow.

I was happy with just getting by and thought I would just make my way into college without really trying. However, life was about to give me a wakeup call.

My cousin and I reached the station, and we were waiting for the train. He was one of the few people who had a smartphone back then. While we waited for the train, my cousin offered to check how much I scored on his iPhone. He punched in my details and waited for the result to load. Seconds seemed like ages. His excited face was replaced with a look of sympathy.

The look on his face only told me that I didn't even get the bare minimum required to get into college. I grabbed his phone and saw how much I scored. My face turned pale.

I got the worst scores possible in math. It would take a miracle for me to go to college.

I later learnt that all my friends did well. My girlfriend did so well that she was going to study medicine. She broke up with me soon after. In my own eyes, everyone around me seemed a lot smarter than me.

I started loathing myself. I lost all self-esteem. Needless to say this was one of the lowest points of my life. I had hit rock bottom, and I didn't know if I even had a future. I would spend the next few months in depression crying about everything that went wrong in my life.

My parents were anxious. However, word got around, and one of my father's friends offered to give me a recommendation. Life gave me another chance, and I somehow went to college.

But this wasn't enough to restore my lost self-esteem. Everyone I knew, got better grades than me. To my young mind, grades were a measure of intelligence. By that measure, I saw myself as having one of the lowest IQs in the world.

The late president of my country once said, "Dreams are not what you have when you sleep. The true dreams are the ones that don't let you sleep."

Becoming the smartest possible version of myself became an obsession that wouldn't let me sleep. I had to become incredibly intelligent, to restore my lost self-esteem. This made me want to learn everything I possibly could, that would push the boundaries of my intelligence.

I decided to start this journey from a place I hated the most – MATH.

It made a lot of sense to me back then to start here because my lack of love for it almost cost me my future. So the first thing I did was learn how to do math in my head.

I read every possible book there was. Tracked down every article and video published on the subject and absorbed it. I combined what I learnt from different places to create my unique system that would let anybody do mental math lightning fast.

The techniques can be learnt in a few minutes. When I first discovered it, it started to change the way I thought.

Since calculations were done in the head, I acquired better memory habits. My concentration and my ability to think improved. But most importantly, learning how to do mental math pushed the boundaries of what I thought was possible.

If I could now do math faster in my head after having hated it for a good part of my life, then I could do anything.

I have been relentlessly pushing the boundaries of my intelligence in every way possible ever since. Now I have an MS degree and work in the field of data analytics.

Math is what I enjoy doing the most. Nobody believes me when I tell them I was an average student in school. They think I was gifted from birth.

Remember, my brain is no different from yours when it comes to numbers. I am just a guy who happens to have a system which makes it easier to do mental math.

This is a system that allows people like us to experience a life where math and numbers become useful, as we experience a new and powerful way to think.

This is the same system that has allowed me to effortlessly calculate numbers in my head faster than a calculator after having hated math for a good part of my life.

And today, in this book, I am going to share the secret of doing mental math with you. The secret that will change your life the way it did mine.

Chapter 1 - What Is Ofpad Math System?

The Ofpad Mental Math system is the only system to do math in your head faster than a calculator making math your new superpower. These radically simple techniques for mental math will work for you even if you hate math and if you are terrible at math, to begin with.

This mental math book was written after analysing the techniques used by over 27 mental math geniuses across the world, from countries including Sweden, India and China. I uncovered these hidden patterns and unique (sometimes 'odd') tactics they use to do math with superhuman speed. And now, starting today, their success can be your success.

What is remarkable about this system is that it is so simple that even a child can do it. The strategies of these math geniuses can now be duplicated easily by you, giving you the power of using math and numbers in your life. You will enjoy an improved memory and develop a laser-sharp concentration, all while you calculate faster than you ever dreamed possible.

You will not find complicated rules that work for specific situations in this book. Instead, you will learn broad concepts that you can apply to all types of math problems.

You will not find math techniques that will require you to use a paper. After you finish this book, you will start calculating in your head faster than before.

However, for the techniques to become second nature to you, you will have to do the practice exercises that come with the book.

Now, before we go any further, I need to be 100% honest with you. If what you want is a "magic pill" that you can swallow to become a genius (something you already know will never work in practice), then you can stop reading this book right now.

The Ofpad Mental Math System isn't that kind of fantasy land nonsense. It's the real deal, and it is only for folks like you who are serious about improving their intelligence, and who are willing to put in the time to practice to hone this new skill.

While doing math in your head isn't "easy" (nothing worthwhile is). With this book, I have made mental math as easy as it can get.

So, if you're like most folks reading this book, and if you're 100% ready to learn the real secrets of doing mental math faster than a calculator, while increasing concentration, developing better memory habits and learning new ways to think and do math entirely in your head faster than a calculator, then the Ofpad Mental Math System isn't just "A" solution for you. It's the "ONLY" solution for you!

You may think you have seen it all when it comes to math. But trust me you would have never seen anything like this before. This system is entirely different, and it's nothing like what you have been taught in school. The method I am about to share with you will shock you, and you will be left wondering why nobody ever taught you this when you were young.

Before we get into the actual techniques, I want to tell you about three things.

If you have a question or hit a road block on your journey to master mental math, we don't want to leave you hanging. We have built a community in Ofpad.com where you can sign-up **for free** and ask your questions. It doesn't matter how silly you think your question is, just ask it in **ofpad.com/mathqa** and we will make sure we answer them. The "Mental Math Q & A Community" chapter will help you sign up to Ofpad and get started in the community, so you can start asking your questions.

The second thing I want to tell you about is learning modalities. People purchase books that they never finish & if they finish it, don't retain a lot. In spite of all the good intentions, it happens because reading is not their learning modality. I want you to not only learn mental math but also master it. We have created a video course for visual and auditory learners to teach mental math techniques using animations. When they do the math later, they visualise the mental process the same way they see in the video. The entire video course can be watched in a couple of hours, so most people don't abandon the course half way, like how they do with the book. If you are someone who never finishes a book or if you are someone who learns better when the information is conveyed in visual format, then check out the video course at **ofpad.com/mathcourse**. If reading is your learning modality then you don't need the video course and this book is everything you need to master mental math.

Irrespective of the format you choose to learn mental math, remember you are getting the same content. Don't get the video course expecting content different from the book because you will be disappointed. Pick the medium that best fits you. What is more important is that you **MASTER** mental math.

The third thing I wanted to mention is practice. Each chapter has an exercise in the end that you must use to practice if you want to master mental math. You can download the rich text PDFs for these exercises from **ofpad.com/mathexercises**. There is also some bonus email practice that is free for you. Once you finish the book, you will find information to get access to this in the chapter on "Eliminating Skill Atrophy".

I hope you enjoy learning how to do Mental Math and we can't wait to hear about your success in the Ofpad community.

Chapter 2 - DS Method of Checking

In this chapter, we will cover the DS Method to check your calculation. DS stands for Digit Sum.

This method has been known to mathematicians for several centuries. But it is not widely known or taught in school, which is why it is not used much in everyday life.

You will be able to check the answers of your addition, subtraction, multiplication and division problems quickly using this DS Method. We are covering this first so that you can use it throughout this book.

Rules To Calculate Digit Sum

Rule 1 - Digit sum is simply the sum of all digits in a number.

For example, the digit sum of 213 is 6. You get this by adding 2 + 1 + 3 which gives us 6.

Number: 213
Digit Sum: 2 + 1 + 3 = 6

Rule 2 - Digit sum should be a single digit number.

For example, the digit sum of 2134 is 1. You get this by adding 2 + 1 + 3 + 4 which gives us 10. Since a digit sum is a single number we must also add 1 + 0 to get 1.

Number: 2134
Digit Sum: 2 + 1 + 3 + 4 = 10
Single Digit: 1 + 0 = 1

Rule 3 - You should ignore the number nine when adding across.

> **Number: 909**
> **Digit Sum: 0 (Ignoring 9s we get 0)**

Rule 4 - You can also ignore the digits which add up to 9 (e.g. 1 & 8 or 3, 4 & 2). Your digit sum will be the same even if you don't ignore the numbers that add up to 9.

In 1802, dropping 1 & 8 which adds up to 9 we get the digit sum as 2.

> **Number: 1802**
> **Digit Sum: 0 + 2 = 2**

Alternatively, without dropping 1 & 8 which adds up to 9, we will still get the same digit sum 2.

> **Number: 1802**
> **Digit Sum: 1 + 8 + 0 + 2 = 11**
> **Single Digit: 1 + 1 = 2**

So your digit sum remains the same irrespective of whether or not you drop the numbers that adds up to 9.

It is okay to drop 9 because digit sum will remain unchanged when you add any number to 9. If you see the additions below, the number you add to 9 is the digit sum of the answer.

$$1 + 9 = 10 = 1 + 0 = \mathbf{1}$$
$$2 + 9 = 11 = 1 + 1 = \mathbf{2}$$
$$3 + 9 = 12 = 1 + 2 = \mathbf{3}$$
$$4 + 9 = 13 = 1 + 3 = \mathbf{4}$$
$$5 + 9 = 14 = 1 + 4 = \mathbf{5}$$
$$6 + 9 = 15 = 1 + 5 = \mathbf{6}$$
$$7 + 9 = 16 = 1 + 6 = \mathbf{7}$$
$$8 + 9 = 17 = 1 + 7 = \mathbf{8}$$
$$9 + 9 = 18 = 1 + 8 = \mathbf{9}$$

Look at the numbers in **bold** in the far left and the far right. They are the same. So by dropping 9 or numbers that add up to 9, you will skip the intermediate addition, which is an unnecessary step during your calculation of the digit sum.

Rule 5 - Decimals work the same way as normal numbers. The decimal point will not affect the digit sum.

Rules To Apply DS Method of Checking

The rule for checking your answers with the DS Method is simple:

1. Whatever you do with the numbers (add, subtract, multiply or divide), also do to the digit sum of the numbers.
2. The result you get from the digit sum of the numbers should be equal to the digit sum of the answer.

Let us look at an example to understand the method.

Checking Addition

Let us check if the sum of 93 + 11 = 104 without explicitly calculating it.

Whatever you do with the numbers, also do to the digit sum of the numbers.

Calculation: 93 + 11 = 104
Digit Sum: 3 + 2 = 5

Dropping 9 from 93 we get the digit sum of 93 as 3.

Adding the two digits of 11, we get the digit sum of 11 as 2.

Adding 3 + 2 we get 5.

The result of the addition is correct because 5 is also the digit sum of the answer 104 (got by adding 1 + 0 + 4 = 5).

Checking Subtraction

Let us check if the difference between 93 - 11 = 82 without explicitly calculating it.

Whatever you do with the numbers, also do to the digit sum of the numbers.

Calculation: 93 - 11 = 82
Digit Sum: 3 - 2 = 1

So dropping 9 from 93 we get the digit sum of 93 as 3.

Adding the two digits of 11, we get the digit sum of 11 as 2.

Subtracting 3 - 2 we get 1.

The result of the subtraction is correct because the digit sum of the answer 82 is also 1 (got by adding 8 + 2 = 10 = 1 + 0 = 1).

Negative Digit Sum During Subtraction

Sometimes the subtraction of the digit sum results in a negative number.

If the subtraction of the digit sums is negative **add 9 to the negative number**.

Let us look at an example.

Let us check if the difference between 23 - 17 is equal to 6 without explicitly calculating it.

Whatever you do with the numbers, also do to the digit sum of the numbers.

Calculation: 23 - 17 = 6
Digit Sum: 5 - 8 = -3 (+9) = 6

So adding 2 and 3 in 23 we get the digit sum of 23 as 5.

Adding the two digits of 17, we get the digit sum of 17 as 8.

Subtracting 5 - 8 we get -3.

If the subtraction of the digit sums is negative, add 9 to the negative number.

So adding 9 to -3 we get 6.

The digit sum of the answer is also 6, so the subtraction is done correctly.

The other way you can check the subtraction using the DS method is by converting it into an addition problem by moving the negative number to the other side of the equal sign. Converting the subtraction problem into an addition problem this way will eliminate the need to add 9 when the digit sums becomes negative.

Let us look at the last example again.

Let us check if the difference between 23 - 17 is equal to 6 without explicitly calculating it.

Calculation: 23 - 17 = 6

If we move the negative number -17 to the other side of the equal sign we get 23 = 6 + 17.

Calculation: 23 = 6 + 17

Now find the digit sum of the numbers.

So adding 2 and 3 in 23 we get the digit sum of 23 as 5.

Calculation: 23 = 6 + 17
Digit Sum: 5

The digit sum of 6 is 6.

Calculation: 23 = 6 + 17
Digit Sum: 5 = 6 +

Adding 1 and 7 in 17 we get the digit sum of 17 as 8.

Calculation: 23 = 6 + 17
Digit Sum: 5 = 6 + 8

Whatever you do to the numbers, also do to the digit sum of the numbers.

Since we are adding 17 and 6 to get 23, we will add 6 + 8 to get 14.

$$\text{Calculation: } 23 = 6 + 17$$
$$\text{Digit Sum: } 5 = 14$$

Since there are two digits in 14 we add 1 + 4 to get the digit sum as 5.

$$\text{Calculation: } 23 = 6 + 17$$
$$\text{Digit Sum: } 5 = 5$$

The digit sum on the other side of the equal sign is also 5, so the subtraction of 23 − 17 is done correctly.

Checking Multiplication

Let us check if 93 multiplied by 11 is equal to 1023 without explicitly calculating it.

Whatever you do with the numbers, also do to the digit sum of the numbers.

$$\text{Calculation: } 93 \times 11 = 1023$$
$$\text{Digit Sum: } 3 \times 2 = 6$$

So dropping 9 from 93 we get the digit sum of 93 as 3.

Adding the two digits of 11, we get the digit sum of 11 as 2.

Multiplying 3 by 2 we get 6.

The result of the multiplication is correct because 6 is also the digit sum of the answer 1023 (got by adding 1 + 0 + 2 + 3 = 6).

Checking Division

Let us check if the division of 110 by 11 is equal to 10 without explicitly calculating it.

Whatever you do with the numbers, also do to the digit sum of the numbers.

Calculation: 110 / 11 = 10
Digit Sum: 2 / 2 = 1

Adding the three digits of 110 we get the digit sum of 110 as 2.

Adding the two digits of 11, we get the digit sum of 11 as 2.

Dividing 2 by 2 we get 1.

The digit sum of the answer 10 is also 1 (got by adding 1 + 0 = 1), so the division is done correctly.

The last example was straight forward, but it will not always work.

Check if 16.1 divided by 7 gives us 2.3.

Calculation: 16.1 / 7 = 2.3

Whatever you do to the numbers, also do to the digit sum of the numbers.

Adding the three digits of 16.1 we get the digit sum of 16.1 as 8.

Calculation: 16.1 / 7 = 2.3
Digit Sum: 8

The digit sum of 7 is 7.

Calculation: 16.1 / 7 = 2.3
Digit Sum: 8   / 7 =

Adding the two digits of 2.3 we get the digit sum of 2.3 as 5.

Calculation: 16.1 / 7 = 2.3
Digit Sum: 8   / 7 = 5

Now we are faced with the problem of dividing 8 by 7 to get 5.

So instead of taking the digit sum and dividing the numbers, convert the division into a multiplication problem.

So the division can be rewritten as 16.1 = 2.3 x 7.

Calculation: 16.1 = 2.3 x 7

If you do the same transformation to the digit sum, you will get 8 = 5 x 7

Calculation: 16.1 = 2.3 x 7
Digit Sum: 8   =   5 x 7

Multiplying 5 and 7 we get 35.

Calculation: 16.1 = 2.3 x 7
Digit Sum: 8   =   35

Taking the digit sum of 35 we get 8.

Calculation: 16.1 = 2.3 x 7
Digit Sum: 8   =   8

Since 8 is also the digit sum in the other side of the equation, we know the division is done correctly.

If the division results in a fixed decimal point, then you can apply the DS method without any manipulation. If the decimal place is repeating, you need to convert the decimal to a remainder.

You then rewrite the problem as Dividend = Quotient x Divisor + Remainder and then calculate the digit sum.

For example, if we divide 47 by 12 we get 3.91666666 (with 6 repeating forever).

Calculation: 47/12 = 3.9166666…

It is not possible to apply the DS method directly here. You must convert the decimal into a remainder.

47 divided by 12 gives us 3 with the remainder 11.

Calculation: 47/12 = 3 r11

You can now check if the division of 47 by 12 gives us the quotient as 3 and the remainder as 11.

You need to rewrite the equation like this Dividend = Quotient x Divisor + Remainder.

That means we write the division as follows:

Calculation: 47 = 3 x 12 + 11

Now find the digit sum of each of the numbers.

Adding the two digits 4 and 7 of 47 we get the digit sum of 47 as 11.

Calculation: 47 = 3 x 12 + 11
Digit Sum: 11

Since there are two digits in 11 we add the 1+1 to get the digit sum as 2.

Calculation: 47 = 3 x 12 + 11
Digit Sum: 2

The digit sum of 3 is 3.

Calculation: 47 = 3 x 12 + 11
Digit Sums: 2 = 3

Next adding the two digits one and two of twelve we get the digit sum of 12 as 3.

Calculation: 47 = 3 x 12 + 11
Digit Sum: 2 = 3 x 3

Adding the two digits of 11 we get the digits sum as 2.

Calculation: 47 = 3 x 12 + 11
Digit Sum: 2 = 3 x 3 + 2

Multiplying 3 with 3 gives us 9.

Calculation: 47 = 3 x 12 + 11
Digit Sum: 2 = 9 + 2

Since it is 9 we drop it and it becomes 0.

$$\text{Calculation: } 47 = 3 \times 12 + 11$$
$$\text{Digit Sum: } 2 = 0 + 2$$

We are left with 2 on either side of the equation so the division is done correctly.

$$\text{Calculation: } 47 = 3 \times 12 + 11$$
$$\text{Digit Sum: } 2 = 2$$

Tip – Reduce Running Total to Single Digits

Reduce the digit sum to a single figure as you go along and don't wait until the end to reduce it to a single digit number.

Suppose you are finding the digit-sum of 512,422.

Start on the left and add across: 5 + 1 + 2 + 4 and so on. Say to yourself only the running totals.

Totals: 6, 8, 12 and reduce to a single figure as you go along. Reduce this 12 to 3 (got by adding 1 + 2).

Go ahead with this 3 and add the remaining two digits of our example:

3, 5, 7. The digit-sum is 7.

This is less difficult than reducing the digit sum to a single digit towards the end. In a very long number, reducing the running total to a single digit number saves you a lot of time.

Limitations of the DS Method

You can only be 90% sure that you have got the correct answer with the DS method.

For example, if you had to check if 14 x 2 = 37, the DS method will say it is correct even though the correct answer is 28.

There will always be one in ten chance that the wrong answer's digit sum matches with the correct answer's digit sum.

Also if you place the decimal point in the wrong place, the DS method won't be able to catch the mistake. For example, if you calculated 16.1 divided by 7 as 23.0 instead of the correct answer 2.3 then the DS method won't be able to identify the mistake.

However, that doesn't make the DS method any less useful.

We will look at how to apply the alternative DD method next. When used in conjunction with the DS method, it will help you check your answers with greater accuracy.

Summary

The DS method of checking will be useful in the coming chapters. It can also be used in any practice problems that you do. Practice will make using these techniques effortless and easy.

If you have any questions, you can ask it in the math Q & A section of the community by going to **ofpad.com/mathqa** and we will make sure to respond. If you have not become a member of the community yet, watch the video **ofpad.com/communityguide** to get started.

If you have enjoyed the book so far, do leave a review on Amazon by visiting **ofpad.com/mathbook**.

If you did not enjoy the book so far, and if you have any general suggestions to improve the book or specific feedback for this chapter, do let us know at **ofpad.com/feedback**. If your feedback helps us improve the book even in small ways, we will thank you by sending a free review copy of our next product when it becomes available.

Chapter 3 - DD Method of Checking

DD stands for Digit Difference. The DD method is another useful way to check your answer because it overcomes most of the limitations of the DS Method. The DD method can also be used to check if the decimal point is placed in the correct position.

Rule To Apply DD Method of Checking

Below is the rule to check your answer with the Digit Difference method:

Step 1 – Add the numbers that are odd starting from the first digit left of the decimal point and then add all the other digits that are in the even place.

Step 2 – Subtract the sum of the digits in the even place from the sum of the digits in the odd place to get the digit difference.

Step 3 – If the subtraction results in a negative number, add 11 to the negative number to get the digit difference.

Step 4 – What you do to the numbers, also do to the digit difference of the numbers.

Remember that just like the digit sum, the digit difference should be a single digit number. So if you get a two digit number during your calculation, make it a single digit number by calculating its digit difference.

Odd & Even Digits in Digit Difference

Let us find the odd and even digits for a few numbers.

Find the odd and even digits of 46

The first odd digit is left of the decimal point. So that makes 6 odd and 4 even.

Number: 46
Odd Place	Even Place
6	4

Find the odd and even digits of 412

The first odd digit is left of the decimal point. So that makes 2 and 4 the odd digits and 1 the even digit.

Number: 412
Odd Place	Even Place
2, 4	1

Find the odd and even digits of 71.23

The first odd digit is left of the decimal point, which would make the first digit right of the decimal point even. So that makes 1 and 3 the odd digits and 7 and 2 the even digits.

Number: 71.23
Odd Place	Even Place
1, 3	7, 2

Find the odd and even digits of 5.67

The first odd digit is left of the decimal point, which would make the first digit right of the decimal point even. So that makes 5 and 7 the odd digits and 6 the even digit.

Number: 5.67
Odd Place	Even Place
5, 7	6

Finding out the odd and even digits is important for the calculation of the digit difference.

Checking Addition

Let us check if 16 + 14 = 30

Calculation: 16 + 14 = 30

The first number is 16.

Step 1 - Add the numbers that are odd starting from the first digit left of the decimal point and then add all the other digits that are in the even place.

In 16, the number 6 is in the odd place from the left and the number 1 is in the even place from the left. Since there is only one-digit in the odd and even place, we have nothing to add.

Number: 16
Odd Place	Even Place
6	1

Step 2 - Subtract the sum of the digits in the even place from the sum of the digits in the odd place.

Subtracting 6 - 1 we get 5 as the Digit Difference of 16.

Calculation: 16 + 14 = 30
Digit Difference: 5

The next number is 14.

Step 1 - Add the numbers that are odd starting from the first digit left of the decimal point and then add all the other digits that are in the even place.

In 14, the number 4 is in the odd place from the left and the number 1 is in the even place from the left. Since there is only one-digit in the odd and even place, we have nothing to add.

Number: 14
Odd Place	Even Place
4	1

Step 2 - Subtract the sum of the digits in the even place from the sum of the digits in the odd place.

Subtracting 4 - 1 we get 3 as the Digit Difference of 14.

Calculation: 16 + 14 = 30
Digit Difference: 5 + 3

The next number is 30.

Step 1 - Add the numbers that are odd starting from the first digit left of the decimal point and then add all the other digits that are in the even place.

0 is in the odd place from the left and 3 is in the even place from the left. Since there is only one-digit in the odd and even place, we have nothing to add.

Number: 30

Odd Place   **Even Place**

0   3

Step 2 - Subtract the sum of the digits in the even place from the sum of the digits in the odd place.

Subtracting 0 - 3 we get -3.

Step 3 - If the subtraction results in a negative number, add 11.

So, we add 11 + -3 to get 8 which is the Digit Difference of 30.

Calculation: 16 + 14 = 30
Digit Difference: 5 + 3 = 8

Step 4 - What you do to the numbers, also do to the digit difference of the numbers.

So adding 5 + 3 we get 8.

Calculation: 16 + 14 = 30
Digit Difference: 5 + 3 = 8

The calculation is correct since 8 is the digit difference of 30 which is on the other side of the equation.

Checking Subtraction

Let us check if 16 - 14 = 2

Calculation: 16 - 14 = 2

When you have to check subtraction, convert the calculation into an addition problem by moving the negative number to the other side of the equation.

So, moving the negative number -14 to the other side of the equation, makes it +14.

The equation becomes 16 = 2 + 14.

Calculation: 16 = 2 + 14

The first number is 16.

Step 1 - Add the numbers that are odd starting from the first digit left of the decimal point and then add all the other digits that are in the even place.

In 16, the number 6 is in the odd place from the left and the number 1 is in the even place from the left. Since there is only one-digit in the odd and even place, we have nothing to add.

Number: 16

Odd Place	Even Place
6	1

Step 2 - Subtract the sum of the digits in the even place from the sum of the digits in the odd place.

Subtracting 6 - 1 we get 5 as the Digit Difference of 16.

Calculation: 16 = 2 + 14
Digit Difference: 5

The next number is 2. Since 2 is a single digit number, the digit difference is going to be the number itself, which is 2 in this case. If you want to calculate the digit difference, you can imagine 2 as 02. Let us see how it works.

Step 1 - Add the numbers that are odd starting from the first digit left of the decimal point and then add all the other digits that are in the even place.

2 is in the odd place from the left and 0 is in the even place from the left. Since there is only one-digit in the odd and even place, we have nothing to add.

Number: 02
Odd Place	Even Place
2	0

Step 2 - Subtract the sum of the digits in the even place from the sum of the digits in the odd place.

Subtracting 2 - 0 we get 2 which is the Digit Difference of 2.

Calculation: 16 = 2 + 14
Digit Difference: 5   2

The next number is 14.

Step 1 - Add the numbers that are odd starting from the first digit left of the decimal point and then add all the other digits that are in the even place.

In 14, the number 4 is in the odd place from the left and the number 1 is in the even place from the left. Since there is only one-digit in the odd and even place, we have nothing to add.

Number: 14

Odd Place	Even Place
4	1

Step 2 - Subtract the sum of the digits in the even place from the sum of the digits in the odd place.

Subtracting 4 - 1 we get 3 as the Digit Difference of 14.

Calculation: 16 = 2 + 14
Digit Difference: 5   2   3

What you do to the numbers, also do to the digit difference of the numbers.

So, adding 2 + 3 we get 5.

The calculation is correct since 5 is also the digit difference of 16 which is on the other side of the equation.

Checking Multiplication

Let us check if 16 x 14 = 224

Calculation: 16 x 14 = 224

The first number is 16.

Step 1 - Add the numbers that are odd starting from the first digit left of the decimal point and then add all the other digits that are in the even place.

In 16, the number 6 is in the odd place from the left and the number 1 is in the even place from the left. Since there is only one-digit in the odd and even place, we have nothing to add.

Number: 16
Odd Place	Even Place
6	1

Step 2 - Subtract the sum of the digits in the even place from the sum of the digits in the odd place.

Subtracting 6 - 1 we get 5 as the Digit Difference of 16.

Calculation: 16 x 14 = 224
Digit Difference: 5

The next number is 14.

Step 1 - Add the numbers that are odd starting from the first digit left of the decimal point and then add all the other digits that are in the even place.

In 14, the number 4 is in the odd place from the left and the number 1 is in the even place from the left. Since there is only one-digit in the odd and even place, we have nothing to add.

Number: 14
Odd Place	Even Place
4	1

Step 2 - Subtract the sum of the digits in the even place from the sum of the digits in the odd place.

Subtracting 4 and 1 we get 3 as the Digit Difference of 14.

Calculation: 16 x 14 = 224
Digit Difference: 5   3

The next number is 224.

Step 1 - Add the numbers that are odd starting from the first digit left of the decimal point and then add all the other digits that are in the even place.

In 224, the numbers 4 and 2 are in the odd place from the left. Adding 4+2 we get 6. In 224, the middle 2 is in the even place from the left, since it is only one number, we have nothing to add.

Number: 224
Odd Place   **Even Place**
4+2   2

Step 2 - Subtract the sum of the digits in the even place from the sum of the digits in the odd place.

Subtracting 6 - 2 we get 4 as the Digit Difference of 224.

Calculation: 16 x 14 = 224
Digit Difference: 5   3   4

Step 4 - What you do to the numbers, also do to the digit difference of the numbers.

Multiplying 5 and 3 we get 15.

Calculation: 16 x 14 = 224
Digit Difference: 5 x 3 = 4(15)

Since 15 is a two-digit number, we should find its digit difference before we can check the answer.

Let us now find the digit difference of 15.

Step 1 - Add the numbers that are odd starting from the first digit left of the decimal point and then add all the other digits that are in the even place.

In 15, the number 5 is in the odd place from the left and the number 1 is in the even place from the left. Since there is only one-digit in the odd and even place, we have nothing to add.

Number: 15
Odd Place   **Even Place**
5   1

Step 2 - Subtract the sum of the digits in the even place from the sum of the digits in the odd place.

Subtracting 5 and 1 we get 4 as the digit difference of 15

Calculation: 16 x 14 = 224
Digit Difference: 5 x 3 = 4

The calculation is correct since 4 is also the digit difference of 224.

Checking Division

Check if 51 divided by 12 gives us 4.25

Calculation: 51 ÷ 12 = 4.25

When you have to check your division, convert the calculation into a multiplication problem by moving the divisor to the other side of the equation.

So, after moving the divisor 12 to the other side of the equal sign the equation becomes:

Calculation: 51 = 4.25 x 12

The first number is 51.

Step 1 - Add the numbers that are odd starting from the first digit left of the decimal point and then add all the other digits that are in the even place.

In 51, the number 1 is in the odd place from the left and the number 5 is in the even place from the left. Since there is only one-digit in the odd and even place, we have nothing to add.

Number: 51
Odd Place	Even Place
1	5

Step 2 - Subtract the sum of the digits in the even place from the sum of the digits in the odd place.

Subtracting 1 - 5 we get -4.

Calculation: 51 = 4.25 x 12
Digit Difference: -4

Step 3 - If the subtraction results in a negative number, add 11.

So, adding 11 to -4 we get 7 as the Digit Difference of 51.

Calculation: 51 = 4.25 x 12
Digit Difference: 7

The next number is 4.25.

Step 1 - Add the numbers that are odd starting from the first digit left of the decimal point and then add all the other digits that are in the even place.

In 4.25, the number 4 is in the odd place from the left. If 4 is odd then the number 2 on the right of the decimal is even and the 5 becomes odd. Adding 4 and 5 we get 9 and since 2 is the only even number, we leave it as it is.

Number: 4.25
Odd Place   **Even Place**
4 + 5   2

Step 2 - Subtract the sum of the digits in the even place from the sum of the digits in the odd place.

Subtracting 9 - 2 we get 7 as the Digit Difference of 4.25.

Calculation: 51 = 4.25 x 12
Digit Difference: 7 = 7

The next number is 12.

Step 1 - Add the numbers that are odd starting from the first digit left of the decimal point and then add all the other digits that are in the even place.

In 12, the number 2 is in the odd place from the left and the number 1 is in the even place from the left. Since there is only one-digit in the odd and even place, we have nothing to add.

Number: 12
Odd Place   **Even Place**
2   1

Subtract the sum of the digits in the even place from the sum of the digits in the odd place.

Subtracting 2 - 1 we get 1.

$$\text{Calculation: } 51 = 4.25 \times 12$$
$$\text{Digit Difference: } 7 = 7 \times 1$$

What you do to the numbers, also do to the digit difference of the numbers.

Since we are multiplying 4.25 x 12, we also multiply 7 and 1 to get 7.

The calculation is correct since 7 is the digit difference of 51 which is on the other side of the equation.

DS Method VS DD Method

Why would you choose the DD method when you have the DS Method?

When we divided 51 by 12 if instead of 4.25 we wrote down the answer 42.5, then the DS method would have missed the mistake, but the DD method would have caught it.

When we multiplied 16 x 14 if instead of 224 we wrote down 242, then the DS method would have missed the mistake, but the DD method would have caught it.

So, use both the DS Method and the DD method together to check your answers with greater accuracy.

Practice will make using these techniques effortless and easy. If you have any questions, you can ask it in the math Q & A section of the community by going to **ofpad.com/mathqa** and we will make sure to respond. If you have not become a member of the community yet, watch the video **ofpad.com/communityguide** to get started.

If you have enjoyed the book so far, do leave a review on Amazon by visiting **ofpad.com/mathbook**.

If you did not enjoy the book so far, and if you have any general suggestions to improve the book or specific feedback for this chapter, do let us know at **ofpad.com/feedback**. If your feedback helps us improve the book even in small ways, we will thank you by sending a free review copy of our next product when it becomes available.

Chapter 4 - Mental Math Q&A Community

We know you are eager to make progress towards mastering mental math. The community might seem like one of those things that you don't need until you need it. So spending some time getting involved with the community is going to pay off big time later on when you get stuck and need help to keep moving forward.

At first glance, taking part in the community might feel a bit less urgent and perhaps even optional. However if you invest just a little time in the community, the rewards will far outweigh your investment in the future.

As you learn to do mental math, you are going to have questions, suggestions or feedback for us. We are very serious about community and support here at Ofpad. Whether it's questions that you have that you want to get answered or if it is just feedback for us to improve, the Ofpad community is a place where you can post it.

A lot of online communities have a bunch of rules and regulations. We are really welcoming at Ofpad and there are not many rules except for these three:

1. No spam or advertising. That is not what this place is for.

2. Be Yourself.

3. Be Kind.

You can search the community and find a lot of interesting conversations, topics, and people by using the search bar at the top. Someone who was in your place in the past might have asked your questions in the Ofpad community. Using the search in the community will help you find answers to those questions quickly.

You might feel a little intimidated or overwhelmed at first. So this section is all about small things you can do within the community, that when combined, will make you feel right at home in the Ofpad community. We made a video guide for getting started with the community, and you can watch it by visiting **ofpad.com/communityguide**. There are three steps to get started with the community, and once you complete these three steps, you will be ready to use the special Mental Math Q & A section to ask any questions you have as you learn to do mental math.

Step 1 - Sign Up or Login

You cannot participate in the community until you become a member of Ofpad. Becoming a member in Ofpad is absolutely **FREE**.

To become a member:

1. Go to Ofpad.com.
2. Click "Sign Up" on the top.
3. Fill out the registration form with your details.
4. Click the submit button.

Our best advice is to be yourself. Using your real name makes you personal, relatable and authentic and that's the stuff relationships are built on in the community.

After you sign up, login:

1. Go to Ofpad.com.
2. Click Login in the top menu.
3. Enter your email address and password that you used when you signed up.
4. Click the Login button.

Step 2 – Complete your Community Profile

After you logged in, you need to complete your community profile.

To complete your profile:

1. Go to **ofpad.com/community**
2. Click on the "Your Profile" button on the top.
3. Click on the "Edit" link on the left
4. Fill in your details to complete your profile
5. Save by clicking the button at the bottom of the page

You can change your profile picture in Gravatar.com by using the email you used in Ofpad. Your picture will be automatically updated in the Ofpad community. Use your real photo so it feels personal. A real photo of you helps us put a face to your name and members will feel they are connecting with the real you.

When you fill your bio in your profile be yourself. Imagine you're talking to someone who's in the same boat as you. Just share something about yourself like you would with a friend.

Step 3 - Introduce Yourself & Welcome Someone Else

The very first thing an Ofpader does is introduce themselves to the community. We have a special section just for introduction in the community. Introducing yourself is probably the most important step in getting involved with the community. It gets you over the "first post" heebie-jeebies.

At Ofpad, we are family and we want to get to know you. In your introduction write:

- what are you working on
- what are your goals
- tell us something interesting or fun we should know about you.

Just do it. Write something about yourself. What is your story? What are your goals and plans if any to make your goals happen? Quickly write something up and hit post. You don't get to think too hard.

Once you hit post, you will have something out there in the community. That wasn't too bad, was it?

Your community profile lists "Topics Started", meaning other members can always refer back to your introduction in the future and learn more about you. Since you'll be super active in the community, this will come in handy!

After you post an intro, reply to another new member's introduction post. The benefits of this community depend on a give/receive model. So start things right by giving back to someone else in the form of a simple welcome reply.

The Ofpad team is always here to help you if you get stuck. You can email us anytime at **hq@ofpad.com**.

Once you complete your Community Initiation, here is the list of things you can do next in the community.

Ask Questions

If you have any question, then the Ofpad community is the place where you can get it answered.

If you have questions, feedback or suggestions related to Mental math, then we have a specific section for this in the community. You can access this section by going to **ofpad.com/mathqa**.

Someone who was in your place in the past might have asked your questions in the Ofpad community. Using the search in the community will help you find answers to those questions quickly.

The mental math book and video course was revised multiple times with new content based on the questions asked in the Q & A section. So, even if you don't have any questions, spend some time exploring the Q & A section for mental math. You might just learn something new.

The more specific, concrete and direct your questions are, the better the feedback from the community will be.

It doesn't matter how silly you think your question is, just ask it. In the Ofpad community, no question is a bad question. So don't be shy. We will be glad to help you out with whatever question you have.

Answering Questions

It's only natural that you might hesitate to add your voice to an existing post, particularly if you're new to this yourself.

However, don't underestimate how valuable your perspective is. Even if you feel you lack experience, hearing the voices of members who are at different levels makes for a diverse and well-rounded conversation. So, don't be shy; be honest, be open and be kind. You know more than you think you do, so try not to over analyse and just dive in.

Head over to the community and see where you can add your voice and help someone out.

If you are having trouble with anything, feel free to email us anytime at **hq@ofpad.com**. We will be happy to help.

Chapter 5 - Introduction to Mental Math

This is an introductory chapter that will tell you how the rest of the book is structured and will also give you a taste of what it is like to do mental maths.

We will cover generic techniques to add, subtract, multiply and divide any set of numbers in the next set of chapters. However, in this chapter, you will first learn a few mental habits you need to do mental math. We will do this by looking at the technique to multiply by 11. The general techniques to multiply will be covered in the coming chapter.

Course Structure

Each chapter in this book will have the following structure:

Section 1 - First, the steps of the mental math technique will be explained.

Section 2 - Then we will apply the fast math technique to 3 different examples to illustrate each step of the method discussed earlier. Each example will have some variation so you will be able to understand how the method is applied completely.

Section 3 - After the three examples which illustrates the mental math method, you will be given practice exercises so that you can apply these techniques yourself and master them. At the end of the book, I will tell you how you can get practice workbooks every week, so you will be able to practice until these techniques become second nature to you.

Throughout this book, there might come a point in time where you might have questions, queries, clarifications, suggestions or feedback. If you have any questions, you can ask it in the math Q & A section of the community by going **ofpad.com/mathqa** and we will make sure to respond. If you have not become a member of the community yet, watch the video **ofpad.com/communityguide** to get started.

Defining a Multiplicand & a Multiplier

Before we get into mental maths techniques, let us quickly define what a multiplicand and a multiplier are.

Take the following example:

43 x 23

43 is the multiplicand. It is the number being multiplied.

23 is the multiplier. It is the number which is multiplying the first number.

The description of the steps in the mental math technique will refer to the numbers in a multiplication as multiplicand and multiplier. So if you are clear on which number is the multiplicand and which number is the multiplier, you will be able to understand the description better. Don't worry if you don't remember it, because the examples illustrating the technique will clear any confusion you might have.

Multiplication By 11

As an introduction to mental math, let us now look at the technique to multiply by 11, along with its three examples illustrating the method.

There are three steps to multiply by 11:

Step 1 - The first number of the multiplicand (number multiplied) is put down as the left-hand number of the answer.

Step 2 - Each successive number of the multiplicand is added to its neighbour to the left.

Step 3 - The last number of the multiplicand becomes the right-hand number of the answer.

Let us look at an example illustrating this method.

Example 1

Let us multiply.

$$\begin{array}{r} 4\,2\,3 \\ \times\,1\,1 \\ \hline \end{array}$$

Step 1 - The first number of the multiplicand (number multiplied) is put down as the left-hand number of the answer.

$$\begin{array}{r} 4\,2\,3 \\ \times\,1\,1 \\ \hline 4\,\_\,\_\,\_ \end{array}$$

So we put down 4 as the left-hand digit of the answer.

Step 2 - Each successive number of the multiplicand is added to its neighbour to the left.

So 4+2 gives us 6, and 6 becomes the second digit of the answer.

$$\begin{array}{r} 4\,2\,3 \\ \times 1\,1 \\ \hline 4\,6\,\_\,\_ \end{array}$$

Moving on to the next two digits, we add 2 + 3 which gives us 5, and 5 becomes the next digit of the answer.

$$\begin{array}{r} 4\,2\,3 \\ \times 1\,1 \\ \hline 4\,6\,5\,\_ \end{array}$$

Step 3 - The last number of the multiplicand becomes the right-hand number of the answer.

So we put down 3 as the last digit of the answer.

$$\begin{array}{r} 4\,2\,3 \\ \times 1\,1 \\ \hline 4\,6\,5\,3 \end{array}$$

The answer is 4653.

Example 2

Let us try another example. Multiply the following:

```
  5 3 4
x 1 1
-------
```

Step 1 - The first number of the multiplicand is put down as the left-hand number of the answer.

So we put down 5 as the left-hand digit of the answer.

```
  5 3 4
x 1 1
-------
5 _ _ _
```

Step 2 - Each successive number of the multiplicand is added to its neighbour to the left.

So 5+3 gives us 8, and 8 becomes the second digit of the answer.

```
  5 3 4
x 1 1
-------
5 8 _ _
```

Moving to the next two digits, we add 3 + 4 which gives us 7, and 7 becomes the third digit of the answer.

```
  5 3 4
x 1 1
-------
5 8 7 _
```

Step 3 - The last number of the multiplicand becomes the right-hand number of the answer.

So we put down 4 as the last digit of the answer.

$$\begin{array}{r}534\\ \times 11\\ \hline 5874\end{array}$$

The answer is 5874.

Example 3

Let us try another example. Multiply 726 by 11.

$$\begin{array}{r}726\\ \times 11\\ \hline ----\end{array}$$

Take a second to apply the technique by yourself as fast as you can. Once you have the answer, you can check the steps below to see if you got your answer right.

Step 1 - The first number of the multiplicand is put down as the left-hand number of the answer.

So we put down 7 as the left-hand digit of the answer.

$$\begin{array}{r}726\\ \times 11\\ \hline 7\_\_\_\end{array}$$

Step 2 - Each successive number of the multiplicand is added to its neighbour to the left.

So 7+2 gives us 9, and 9 becomes the second digit of the answer.

```
    7 2 6
   x 1 1
   ─────
   7 9 _ _
```

Moving to the next two digits, we add 2 + 6 which gives us 8, and 8 becomes the third digit of the answer.

```
    7 2 6
   x 1 1
   ─────
   7 9 8 _
```

Step 3 - The last number of the multiplicand becomes the right-hand number of the answer.

So we put down 6 as the last digit of the answer, and the final answer is 7986.

```
    7 2 6
   x 1 1
   ─────
   7 9 8 6
```

If you got your answer wrong, don't worry. Just revisit the techniques and examples we covered in this chapter.

Carrying Over In Mental Math

Now when you add the numbers, and if it results in a sum which has two digits, you will have to carry over the first digit.

So here is the complete rule for multiplication by 11 incorporating the step of carrying over.

Step 1 - The first number of the multiplicand is put down as the left-hand number of the answer.

Step 2 - Each successive number of the multiplicand is added to its neighbour to the left.

Step 3 - If the addition results in two figures, carry over the 1 (Note: The two figure number will never be more than 18 (got by adding 9+9)).

Step 4 - The last number of the multiplicand becomes the right-hand number of the answer.

Note: Only the third step is new. The rest remains the same.

Example 1

Let us look at an example. Multiply 619 by 11.

$$\begin{array}{r} 6\ 1\ 9 \\ \times 1\ 1 \\ \hline \_\ \_\ \_\ \_ \end{array}$$

Step 1 - The first number of the multiplicand is put down as the left-hand number of the answer.

So we put down 6 as the left-hand digit of the answer.

$$\begin{array}{r} 6\ 1\ 9 \\ \times 1\ 1 \\ \hline 6\ \_\ \_\ \_ \end{array}$$

Step 2 - Each successive number of the multiplicand is added to its neighbour to the left.

So 6+1 gives us 7, and 7 becomes the second digit of the answer.

```
  619
x  11
6 7 _ _
```

Moving to the next two digits, we add 1 + 9 which gives us 10.

```
  6 1 9
x   1 1
  6 7 _ _
    1 0
```

Step 3 - If the addition results in two figures, carry over the 1.

The second digit of the sum, which is 0 becomes the third digit of the answer.

```
  6 1 9
x   1 1
  6 7 0 _
    1
```

Since the addition resulted in two digits, we carry over the one, so the 7 now becomes 8.

```
  6 1 9
x   1 1
  6 8 0 _
```

Step 4 - The last number of the multiplicand becomes the right-hand number of the answer.

So we put down 9 as the last digit of the answer. The final answer is 6809.

$$\begin{array}{r} 6\,1\,9 \\ \times\,1\,1 \\ \hline 6\,8\,0\,9 \end{array}$$

This concept of carrying over applies to all mental math techniques that we will see in the future chapters. We are looking at this as a separate step in this first chapter for understanding and clarity. However, in the later chapters, we will skip describing this as a separate step.

Example 2

Let us look at another example. Multiply 348 x 11.

$$\begin{array}{r} 3\,4\,8 \\ \times\,1\,1 \\ \hline \_\,\_\,\_\,\_ \end{array}$$

Step 1 - The first number of the multiplicand is put down as the left-hand number of the answer.

So we put down 3 as the left-hand digit of the answer.

$$\begin{array}{r} 3\,4\,8 \\ \times\,1\,1 \\ \hline 3\,\_\,\_\,\_ \end{array}$$

Step 2 - Each successive number of the multiplicand is added to its neighbour to the left.

So 3+4 gives us 7. This becomes the second digit of the answer.

```
  3 4 8
  x 1 1
  ─────
  3 7 _ _
```

Moving to the next two digits, we add 4 + 8 which gives us 12.

```
  3 4 8
  x 1 1
  ─────
  3 7 _ _
    1 2
```

Step 3 - If the addition results in two figures, carry over the 1.

The second digit of the sum, which is 2 becomes the third digit of the answer.

```
  3 4 8
  x 1 1
  ─────
  3 7 2 _
      1
```

Since the addition resulted in two digits, we carry over the one, so the 7 now becomes 8.

```
  3 4 8
  x 1 1
  ─────
  3 8 2 _
```

Step 4 - The last number of the multiplicand becomes the right-hand number of the answer.

So we put down 8 as the last digit of the answer.

$$\begin{array}{r} 3\,4\,8 \\ \times\,1\,1 \\ \hline 3\,8\,2\,8 \end{array}$$

The answer is 3828.

If carrying numbers over seems to strain your mental faculties a bit, do not worry. Carrying over numbers will become effortless and easy the more your practice and progress through this book. This might probably be the first time you are trying to carry over the numbers in your head, so it will take a bit of getting used to.

Example 3

Let us look at another example. Multiply 428 by 11.

$$\begin{array}{r} 4\,2\,8 \\ \times\,1\,1 \\ \hline \_\,\_\,\_\,\_ \end{array}$$

Take a second to apply the technique by yourself as fast as you can. Once you have the answer, you can check the steps below to see if you got your answer right.

Step 1 - The first number of the multiplicand (number multiplied) is put down as the left-hand number of the answer.

So we put down 4 as the left-hand digit of the answer.

```
  4 2 8
  x 1 1
  ─────
  4 _ _ _
```

Step 2 - Each successive number of the multiplicand is added to its neighbour to the left.

So 4+2 gives us 6. This becomes the second digit of the answer.

```
  4 2 8
  x 1 1
  ─────
  4 6 _ _
```

Moving to the next two digits, we add 2 + 8 which gives us 10.

```
  4 2 8
  x 1 1
  ─────
  4 6 _ _
    1 0
```

Step 3 - If the addition results in two figures, carry over the 1.

The second digit of the sum, which is 0 becomes the third digit of the answer.

```
  4 2 8
  x 1 1
  ─────
  4 6 0 _
      1
```

Since the addition resulted in two digits, we carry over the one. So 6 now becomes 7.

```
  4 2 8
  x 1 1
  ─────
  4 7 0
```

Step 4 - The last number of the multiplicand becomes the right-hand number of the answer.

So we put down 8 as the last digit of the answer.

```
  4 2 8
  x 1 1
  ─────
  4 7 0 8
```

The final answer is 4708.

Read this chapter again if necessary. Then go to the practice section and complete the exercises.

You might have understood the technique, but it will take practice before the technique becomes second nature to you.

If you have any questions, you can ask it in the math Q & A section of the community by going **ofpad.com/mathqa** and we will make sure to respond. If you have not become a member of the community yet, watch the video **ofpad.com/communityguide** to get started.

If you have enjoyed the book so far, do leave a review on Amazon by visiting **ofpad.com/mathbook**.

If you did not enjoy the book so far, and if you have any general suggestions to improve the book or specific feedback for this chapter, do let us know at **ofpad.com/feedback**. If your feedback helps us improve the book even in small ways, we will thank you by sending a free review copy of our next product when it becomes available.

Once you finished the practice workbook, move on to the next chapter.

Exercises

Download the rich PDFs for these exercises from **ofpad.com/mathexercises**.

1)
54
x 11

2) 34
x 11

3) 46
x 11

4) 984
x 11

5) 723
x 11

6) 342
x 11

7) 424
x 11

8) 216
x 11

9) 923
x 11

10) 3594
x 11

11) 9035
x 11

12) 1593
x 11

13) 6770
x 11

14) 5459
x 11

15) 2696
x 11

16) 7537
x 11

17) 4921
x 11

18) 6871
x 11

Answers

1) 594

2) 374

3) 506

4) 10,824
5) 7,953
6) 3,762
7) 4,664
8) 2,376
9) 10,153
10) 39,534
11) 99,385
12) 17,523
13) 74,470
14) 60,049
15) 29,656
16) 82,907
17) 54,131
18) 75,581

Chapter 6 - The Inefficient Way to Do Math

The technique you learnt in the last chapter is fine for multiplication by 11 but what about larger numbers? We will cover that in the next few chapters. You don't have to memorise special rules for every number. A few generic techniques will help you do arithmetic for any pair of numbers.

Before we proceed to these generic mental math techniques, I must tell you about the real problem we face. This is a mistake you have been making that limits the speed with which you do the math.

And it is not your fault. You have just been taught to do this in school.

Many people think they need a calculator or at least a pen and paper to be able to do the math. The real problem is the fact that you have been taught an inefficient way to do the math in school. And this is what makes it difficult for you to do the math, leave alone doing it in your head.

One of the common lies people believe is that you need an aptitude for math to be good at it. You might have heard that before or in the past you may have even believed it yourself. Yet nothing could be further from the truth.

If you are one of the millions who fell victim to believing this lie, then you must decide now to believe the truth instead. Because if you don't, you will never experience the usefulness of math and numbers in your life, even if you are extremely smart, to begin with.

It is not your fault because you have been led to believe this by our math education system. So if you struggle to do math in your head don't blame yourself, blame the system which still teaches students outdated ways to do math. The system is designed in such a way that only 1% of people get it while the rest of us merely get by.

When I reveal the simple secret of doing mental math in the next chapter, you will be shocked why they never taught you this at school. And when you learn it, math will immediately start becoming more useful and enjoyable.

The truth is that the way we were taught math in school only slows us down because it uses too much of your working memory.

Working memory is the short-term memory you need to organise information to complete a task. It is like the RAM of your computer.

The way we were taught to do math in school is so inefficient that it eats up so much of our working memory when we try to do the math in our head. So our brain slows down like an overloaded computer. This is what makes mental math so hard to do.

To do math faster in your head, you must do the opposite of what you have been taught in school. I'll show you exactly how to do this in just a few minutes.

First, I need to show you why you need to avoid doing what you have been taught in school. This might surprise you. Most people think this is the only one way to calculate, but in reality, this is what slows you down.

I will show you what I mean with an example.

Say we wanted to multiply 73201 by 3.

$$\begin{array}{r} 73201 \\ \times\ 3 \\ \hline \end{array}$$

We will start by multiplying one and three to get three. Then we multiply zero and three to get zero. We proceed to multiply two and three to get six. Then we multiply three and three to get nine, and finally, we multiply seven and three to get twenty-one.

This wasn't very hard, and in fact, it would only take most people seconds to multiply the individual numbers.

However, to get the final answer, you need to remember every single digit you calculated so far to put them back together. You might even end up multiplying again because you forgot one of the numbers.

So it takes quite a bit of effort to get the answer of the multiplication because you spend so much time trying to recall the numbers you already calculated.

Math would be a whole lot easier to do in your head if you didn't have to remember so many numbers.

In school, we have been taught to write down the numbers on paper to free up our working memory. There is another way, and I will show you that in a moment when we cover the LR method.

Now imagine when we tried to multiply 73201 x 3, if you could have come up with the answer, in the time it took you to multiply the individual numbers. Wouldn't you have solved the problem faster than the time it would have taken you to punch in the numbers inside a calculator?

I will show you how to do that in the next chapter.

Chapter 7 - Introducing The LR Method

The secret of doing mental math fast in your head is to do the opposite of what you have been taught in school.

This technique is called the LR method.

Let me tell you if you do not do this, any trick you can learn to do mental math will be useless. This secret is like the rocket fuel for your mental math skills.

Everything else just builds on top of this. This simple technique works so well because it frees your working memory almost completely.

LR stands for **L**eft to **R**ight.

The secret to doing mental math is to calculate from left to right instead of from right to left.

This is the opposite of what you have been taught in school.

Let us try to do the earlier example where we multiplied 73201 x 3.

But this time let us multiply from left to right.

```
73201
x   3
------
```

Try to do it yourself before reading further. I bet you will have no trouble calling out the answer to the multiplication problem.

Multiplying 7 x 3 gives us 21.

```
    7 3 2 0 1
        x   3
    2 1 _ _ _ _
```

Multiplying 3 x 3 gives us 9.

```
    7 3 2 0 1
        x   3
    2 1 9 _ _ _
```

Multiplying 3 x 2 gives us 6.

```
    7 3 2 0 1
        x   3
    2 1 9 6 _ _
```

Multiplying 0 x 3 gives us 0.

```
    7 3 2 0 1
        x   3
    2 1 9 6 0 _
```

Multiplying 3 x 1 gives us 3.

```
    7 3 2 0 1
        x   3
    2 1 9 6 0 3
```

Did you notice that you started to call out the answer before you even finished the whole multiplication problem? You don't have to remember a thing to recall and use later. So you end up doing math a lot faster.

You just did this for a one-digit multiplier 3. Imagine calculating with the same speed for more complex numbers like two-digit and three-digit multipliers.

Imagine doing addition, subtraction and division with the same speed.

Let me assure you that it is as ridiculously simple as multiplying using the LR method and this will be covered in detail in the next two chapters.

Before we cover that, walk with me for a moment as we imagine you waking up tomorrow being able to do lightning fast math in your head. Your family and friends are going to look at you like you are some kind of a genius.

Since calculations are done in your head, you would have acquired better memory habits in the process. Your concentration and your ability to think quickly would have also improved. So you will not just look like a genius. You will actually be one.

You know what the best part is? Immediately after completing this book, you will start thinking like a genius. This will, in turn, start to positively influence other areas of your life. I am really excited to show you how.

Chapter 8 - Addition & Subtraction

In this chapter, we will look at how to do addition and subtraction fast using the LR Method. The techniques are similar so we will cover both addition and subtraction in the same chapter.

LR Addition

We saw in the earlier chapter that the secret of mental calculation is to calculate from left to right instead of right to left. When you do this, you will start calling out the answer, before you even complete the full calculation.

Solving math from right to left might be good for pen and paper math. However, when you do math the way you have been taught in school, you will be generating the answer in reverse, and this is what makes math hard to do in your head.

LR method might seem unnatural at first, but you will discover that solving math from left to right, is the most natural way to do calculations in your head.

So let us apply this to addition.

Example 1

Add the following numbers together:

$$\begin{array}{r} 5\,3\,2\,1 \\ +\,1\,2\,3\,4 \\ \hline \end{array}$$

The rule is simple. Add from left to right. One-digit at a time.

So adding 5 + 1 gives us 6.

```
  5 3 2 1
+ 1 2 3 4
─────────
  6 _ _ _
```

Adding 3 + 2 gives us 5.

```
  5 3 2 1
+ 1 2 3 4
─────────
  6 5 _ _
```

Adding 2 + 3 gives us 5.

```
  5 3 2 1
+ 1 2 3 4
─────────
  6 5 5 _
```

Adding 1 + 4 gives us 5. And you have the answer 6555.

```
  5 3 2 1
+ 1 2 3 4
─────────
  6 5 5 5
```

The individual steps were broken down to represent the mental process involved to arrive at the answer. When you do the calculation, it should only take you seconds to arrive at the final answer.

Example 2

Let us try another example.

Add the following numbers together:

```
  9 8 8 1
+ 1 2 3 4
---------
_ _ _ _ _
```

The rule is simple. Add from left to right. One-digit at a time.

So, adding 9 + 1 gives us 10.

```
  9 8 8 1
+ 1 2 3 4
---------
1 0 _ _ _
```

Adding 8 + 2 gives us 10.

```
  9 8 8 1
+ 1 2 3 4
---------
1 0 _ _ _
  1 0
```

Carry over the 1, and the 0 becomes 1.

```
  9 8 8 1
+ 1 2 3 4
---------
1 1 0 _ _
```

Adding 8 + 3 gives us 11.

```
  9 8 8 1
+ 1 2 3 4
---------
1 1 0 _ _
    1 1
```

Carry over the 1, and the 0 becomes 1.

```
  9 8 8 1
+ 1 2 3 4
  1 1 1 1 _
```

Adding 1 + 4 gives us 5

```
  9 8 8 1
+ 1 2 3 4
  1 1 1 1 5
```

And you have the final answer 11,115.

Example 3

Let us try another example.

Add the following numbers together:

```
  8 3 7 2
+ 4 6 3 6
  _ _ _ _ _
```

The rule is simple. Add from left to right. One-digit at a time.

Take a second to apply the technique by yourself as fast as you can. Once you have the answer, you can check the steps below to see if you got your answer right.

Add from left to right. So adding 8 + 4 gives us 12

```
  8 3 7 2
+ 4 6 3 6
  1 2 _ _ _
```

Adding 3 + 6 gives us 9.

```
  8 3 7 2
+ 4 6 3 6
---------
  1 2 9 _ _
```

Adding 7 + 3 gives us 10.

```
  8 3 7 2
+ 4 6 3 6
---------
  1 2 9 _ _
      1 0
```

Carry over the 1, and the 29 becomes 30.

```
  8 3 7 2
+ 4 6 3 6
---------
  1 3 0 0 _
```

Adding 2 + 6 gives us 8.

```
  8 3 7 2
+ 4 6 3 6
---------
  1 3 0 0 8
```

You have the final answer 13,008.

If you got your answer wrong, don't worry. Just revisit the techniques and examples we covered in this chapter.

Tip - When you do the problems in your head, don't just visualise it in your mind, try to hear them as well. For example, when you are adding 8432 + 4636 say eight **thousand** four hundred and thirty-two plus four **thousand** six hundred and thirty-six. You stress the digit you are adding (thousand in this case) to keep track of where you are at.

You can do this for all methods taught in this book. When initially solving the problems, practice the problems by saying it out loud. Saying something out loud will serve as an additional memory aid which clears up more of your working memory.

Rounding Up Before Calculating

A useful technique for addition is to round up the number first, before doing the addition.

$$4529 = 5000 - ?$$

But the problem is finding out how much you rounded-up so you can use it later on.

Take the above example. It is easy to round up 4529 to 5000. However, you need to know how much you rounded-up. Without a technique to find that out, you will end up doing a subtraction of 5000 – 4529 which will defeat the whole purpose of rounding up.

Finding How Much You Rounded-up

To find how much you rounded-up without explicitly doing the subtraction, remember the following:

Step 1 - The last digits add up to 10.

Step 2 - The remaining digits add up to 9.

Let us look at our earlier example:

$$4529 = 5000 - 471$$

In this example, when you round up 4529 to 5000, you have rounded-up by 471.

The last digits of 4529 and 471 add up to 10.

$$452\mathbf{9} = 5000 - 47\mathbf{1}$$
$$\mathbf{9 + 1 = 10}$$

All the other digits 5 and 2 (in 4529) & 4 and 7 (471) add up to 9. So 5 + 4 is 9 & 2 + 7 is again 9.

$$4\mathbf{52}9 = 5000 - \mathbf{47}1$$
$$\mathbf{5 + 4 = 9}$$
$$\mathbf{2 + 7 = 9}$$

Knowing that the last digits add up to 10 and other digits add up to 9, you will be able to arrive at the amount you rounded-up quickly.

Let us do a few rounding up exercises.

Exercise 1

Round up 84791 to 100,000 and find out how much you rounded-up by.

$$\mathbf{84791 = 100{,}000 - \;?}$$

Remember the last digits add up to 10. All the other digits add up to 9.

Take a second to call out the answer.

Then proceed below to check if you got it right.

The answer is 15209.

9 in 1520<u>9</u> adds up with 1 of 8479<u>1</u> to give 10.

All the other digits add up to 9.

Exercise 2

Let us try another number.

Round up 7423 to 10,000 and find out how much you rounded-up by.

$$7423 = 10{,}000 - ?$$

Remember the last digits add up to 10. All the other digits add up to 9.

Take a second to call out the answer.

Then proceed below to check if you got it right.

The answer is 2577.

7 in 257<u>7</u> adds up with 3 of 742<u>3</u> to give 10.

All the other digits add up to 9.

Exercise 3

Let us try another number.

Round up 892 to 1000 and find out how much you rounded-up by.

$$892 = 1000 - ?$$

Take a second to call out the answer.

Then proceed below to check if you got it right.

The answer is 108.

8 in 10<u>8</u> adds up with 2 of 89<u>2</u> to give 10.

All the other digits add up to 9.

Exercise 4

Let us try one last example.

Round up 27 to 100 and find out how much you rounded-up by.

$$27 = 100 - ?$$

Remember the last digits add up to 10. All the other digits add up to 9.

Take a second to call out the answer.

Then proceed below to check if you got it right.

The answer is 73.

3 in 73 adds up with 7 of 27 to give 10.

The other digits, i.e. 2 in 27 and 7 in 73 add up to 9.

LR Addition With Rounding Up

Now that you learnt how to find out how much you rounded-up, let us look at the method to do LR Addition with rounding up. It is extremely simple.

Step 1 - Round up a number.

Step 2 - Add from left to right to the amount you rounded-up.

Step 3 - Subtract the amount you rounded-up from the sum.

Let us look at an example and try to do left to right addition after rounding up one of the numbers.

Example 1

Add the following numbers:

$$\begin{array}{r} 9981 \\ +1234 \\ \hline \end{array}$$

Step 1 - Round up the first number.

$$\begin{array}{r} 9981 = (10000 - 19) \\ +1234 \\ \hline \end{array}$$

So when you round up 9981, you have 10,000, and the amount you rounded-up is 19.

$$\begin{array}{r} 10000 - 19 \\ +1234 \\ \hline \end{array}$$

If you have trouble finding the amount you rounded-up, check the section of rounding up we covered where we saw how the last digits add up to 10 and the remaining digits add up to 9.

Step 2 - Now add from left to right to the amount you rounded-up.

$$\begin{array}{r} 10000 - 19 \\ +1234 \\ \hline 11234 \end{array}$$

So if you add 10,000 with 1234, you get 11234. Rounding up made the addition process easy.

Step 3 - Now subtract the amount you rounded-up from the sum.

$$\begin{array}{r} 11234 \\ -19 \\ \hline \end{array}$$

So subtract 19 from 11234.

Subtract from left to right.

We can put down 112 before proceeding to do the subtraction.

$$\begin{array}{r} 11234 \\ -19 \\ \hline 112\_\_ \end{array}$$

Subtract 3 - 1 to get 2.

```
  1 1 2 3 4
 -      1 9
  ─────────
  1 1 2 2 _
```

Next, you have to subtract 4 – 9 so you have to borrow 1 from the 2. So now the 2 becomes 1.

```
  1 1 2 3 4
 -      1 9
  ─────────
  1 1 2 1 _
```

And 14 – 9 is 5.

```
  1 1 2 3 4
 -      1 9
  ─────────
  1 1 2 1 5
```

11,215 is your final answer.

Note that we deep dived into the steps so that you have clarity and understanding. When you do the steps in your mind, it will take less than 5 seconds to do this entire calculation. Each chapter will explain the technique you should apply, and the practice exercises will take the speed with which you apply the technique to the next level.

Example 2

Let us try another example now.

Add 5492 with 8739.

$$\begin{array}{r}5492\\+8739\\\hline\end{array}$$

Step 1 - Round up the first number.

$$\begin{array}{r}5492 = (5500-8)\\+8739\\\hline\end{array}$$

So when you round up 5492, you have 5500, and the amount you rounded-up is 8. Note that I could just round up to 6000 with the rounded-up value being 508. But rounding up to 5500 will simplify the subtraction step in this example. The smaller the amount you round up, the easier it is to calculate.

$$\begin{array}{r}5500-8\\+8739\\\hline\end{array}$$

If you have trouble finding the amount you rounded-up, check the section on rounding up where we saw how the last digits add up to 10 and the remaining digits add up to 9.

Step 2 - Add from left to right to the amount you rounded-up.

So if you add 5500 with 8739, you get 14239. Rounding up made the addition process a little easier.

$$\begin{array}{r}5500-8\\+8739\\\hline 14239\end{array}$$

Step 3 - Subtract the amount you rounded-up from the sum.

So subtract 8 from 14239. Subtract from left to right.

$$\begin{array}{r} 1\,4\,2\,3\,9 \\ -8 \\ \hline 1\,4\,2\,3\,1 \end{array}$$

Now you have the final answer 14,231.

Example 3

Let us try another example now.

Add 7997 with 5347.

$$\begin{array}{r} 7\,9\,9\,7 \\ +\,5\,3\,4\,7 \\ \hline \end{array}$$

Take a second to apply the technique by yourself as fast as you can. Once you have the answer, you can check the steps below to see if you got your answer right.

Step 1 - Round up the first number.

$$\begin{array}{r} 7\,9\,9\,7 = (8000 - 3) \\ +\,5\,3\,4\,7 \\ \hline \end{array}$$

So when you round up 7997, you have 8,000. The amount you rounded-up is 3.

$$8000-3$$
$$+5347$$
$$\overline{}$$

Step 2 - Add from left to right to the amount you rounded-up.

So if you add 8,000 with 5347, you get 13347. Rounding up made the addition process easy.

$$8000-3$$
$$+5347$$
$$\overline{13347}$$

Step 3 - Subtract the amount you rounded-up from the sum.

So subtract 3 from 13347.

$$13347$$
$$-3$$
$$\overline{13344}$$

Now you have the final answer 13344.

If you got your answer wrong, don't worry. Just revisit the techniques and examples we covered in this chapter.

If you have any questions, you can ask it in the math Q & A section of the community **ofpad.com/mathqa** and we will make sure to respond. If you have not become a member of the community yet, watch the video **ofpad.com/communityguide** to get started.

When Should You Round Up For Addition?

Now you understand how to add and round up, should you always be rounding up before adding or should you do it only during specific situations? Sometimes rounding up simplifies the addition whereas during other times, rounding up just complicates the addition and creates an unnecessary step. So how do you decide when to round up and when not to round up?

The short answer is that **rounding up should only be done when you have to carry over** a lot of numbers.

Let us try to add 9898 + 4343:

$$\begin{array}{r} 9898 \\ +4343 \\ \hline \end{array}$$

You could do this problem with just the LR method without any rounding up. However, if you do that you will have to carry over a number during every step. If you round up the first number, it will make things a lot easier.

You can round up 9898 by 102 to get 10,000.

$$\begin{array}{r} 9898 = (10000 - 102) \\ +4343 \\ \hline \end{array}$$

Adding 10,000 with 4343 gives us 14,343.

```
   10000 - 102
  + 4 3 4 3
  ─────────
    14343
```

Then subtract the amount you rounded-up which is 102 to get 14,241.

```
    14343
  -   102
  ─────────
    14241
```

Try to do add 9898 + 4343 in your head again, with and without rounding up and you will realize you are flexing fewer neurons when you round up.

So this means rounding up is the magic you do before you do every addition right? No.

Let us say you want to add 4343 + 1234.

```
    4343
  + 1234
  ─────────
```

You don't have to carry over any number to do this calculation. So the LR addition WITHOUT rounding up makes sense. But for the fun of it, try to do the same problem by rounding up. You will find that rounding up only increases the mental effort required to do the math.

So remember to only round up when you have to carry over a lot of numbers. If you don't have to carry over any number or if you have to carry over just one-digit, you will end up complicating the problem by rounding up.

Addition Exercises

Download the rich PDFs for these exercises from **ofpad.com/mathexercises**.

```
01) 33        07) 115       13) 2771
   + 20         +596          +3216

02) 77        08) 485       14) 6526
    +97         +327          +3057

03) 82        09) 114       15) 6491
    +63         +164          +2273

04) 157       10) 4942      16) 3878
   +836         +2332         +5483

05) 214       11) 7241      17) 7682
    +155        +9508         +8903

06) 865       12) 8699      18) 9616
    +467        +9897         +9202
```

Addition Answers

01) 53   07) 711   13) 5,987
02) 174   08) 812   14) 9,583
03) 145   09) 278   15) 8,764
04) 993   10) 7,274   16) 9,361
05) 369   11) 16,749   17) 16,585
06) 1,332   12) 18,596   18) 18,818

LR Subtraction

Now we will look at subtraction. The steps are same as that of addition. We briefly saw how to subtract from left to right when we covered rounding up for addition. So we will just jump straight into an example for subtraction.

Example 1

Subtract 1234 from 5389.

```
  5 3 8 9
- 1 2 3 4
---------
  _ _ _ _
```

The rule is simple. Subtract from left to right. One-digit at a time.

So 5 - 1 gives us 4.

```
  5 3 8 9
- 1 2 3 4
---------
  4 _ _ _
```

Subtracting 3 - 2 gives us 1.

```
  5 3 8 9
- 1 2 3 4
---------
  4 1 _ _
```

Subtracting 8 - 3 gives us 5.

```
  5 3 8 9
- 1 2 3 4
---------
  4 1 5 _
```

Subtracting 9 - 4 gives us 5.

```
  5 3 8 9
- 1 2 3 4
---------
  4 1 5 5
```

And you have the answer 4155.

Example 2

Let us try another example. Subtract 5741 from 8431.

```
  8 4 3 1
- 5 7 4 1
---------
  _ _ _ _
```

The rule is simple. Subtract from left to right. One-digit at a time.

So, 8 - 5 gives us 3.

```
  8 4 3 1
- 5 7 4 1
---------
  3 _ _ _
```

To subtract 4 - 7 you will have to borrow 1 from 3. So the 3 becomes 2.

```
  8 4 3 1
- 5 7 4 1
---------
  2 _ _ _
```

Subtracting 14 - 7 gives us 7.

```
  8 4 3 1
- 5 7 4 1
---------
  2 7 _ _
```

To subtract 3 - 4 you will have to borrow 1 from 7. So the 7 becomes 6.

```
  8 4 3 1
- 5 7 4 1
---------
  2 6 _ _
```

Subtracting 13 - 4 gives us 9.

```
  8 4 3 1
- 5 7 4 1
---------
  2 6 9 _
```

Subtracting 1 - 1 gives us 0.

```
  8 4 3 1
- 5 7 4 1
---------
  2 6 9 0
```

The final answer is 2690.

Example 3

Let us try another example. Subtract 3756 from 7328.

```
  7 3 2 8
- 3 7 5 6
---------
  _ _ _ _
```

The rule is simple. Subtract from left to right. One-digit at a time.

Take a second to apply the technique by yourself as fast as you can. Once you have the answer, you can check the steps below to see if you got your answer right.

Subtract from left to right. One-digit at a time. So 7 - 3 gives us 4.

```
  7 3 2 8
- 3 7 5 6
---------
  4 _ _ _
```

To subtract 3 - 7 you will have to borrow 1 from 4. So the 4 becomes 3.

```
  7 3 2 8
- 3 7 5 6
---------
  3 _ _ _
```

Subtracting 13 - 7 now gives us 6.

```
  7 3 2 8
- 3 7 5 6
---------
  3 6 _ _
```

To subtract 2 - 5 you will have to borrow 1 from 6. So the 6 becomes 5.

```
  7 3 2 8
-  3 7 5 6
  ‾‾‾‾‾‾‾
    3 5 _ _
```

Subtracting 12 - 5 gives us 7.

```
  7 3 2 8
-  3 7 5 6
  ‾‾‾‾‾‾‾
    3 5 7 _
```

Subtracting 8 - 6 gives us 2.

```
  7 3 2 8
-  3 7 5 6
  ‾‾‾‾‾‾‾
    3 5 7 2
```

Now you have the final answer 3572.

If you got your answer wrong, don't worry. Just revisit the techniques and examples we covered in this chapter.

If you have any questions, you can ask it in the math Q & A section of the community **ofpad.com/mathqa** and we will make sure to respond. If you have not become a member of the community yet, watch the video **ofpad.com/communityguide** to get started.

The last problem was a little harder than the rest because you had to borrow numbers from the neighbour during two separate steps. That is where the magic of rounding up will come in.

LR Subtraction With Rounding Up

When you did LR Subtraction, you might have noticed that it was a bit tedious to do the subtraction when you had to borrow numbers from the neighbour. This is where rounding up numbers really helps. Rounding up also applies to subtraction. But instead of subtracting the amount you rounded-up (as we did in addition), you will now add the amount you rounded-up.

Rounding up before calculating is usually more useful for subtraction than it is for addition. That is because for most of us it is generally easier to do mental addition than it is to do mental subtraction. Borrowing numbers from the neighbour during subtraction is not as easy as carrying over numbers during addition. When you round up before you subtract, you make the subtraction problem into a simple addition problem.

If required, revisit the section of rounding up where we covered how to find how much you rounded-up. The rule to round up and subtract is simple:

Step 1 - Round up the second number.

Step 2 - Subtract from left to right the amount you rounded-up to.

Step 3 - Add the amount you rounded-up with the subtracted amount.

Let us look at an example.

Example 1

Let us try to do left to right subtraction after rounding up the numbers. Subtract 3898 from 4530.

$$\begin{array}{r} 4530 \\ -\ 3898 \\ \hline \end{array}$$

Notice that this is a hard subtraction because you have to borrow a digit in almost every step. However, this problem is a piece of cake when you round up before you calculate. I will show you how.

Step 1 - Round up the second number.

$$\begin{array}{r} 4530 \\ \underline{-\ 3898} = -(4000 - 102) \\ \hline \end{array}$$

So when you round up 3898, you have 4,000, and the amount you rounded-up is 102.

$$\begin{array}{r} 4530 \\ \underline{-4000} + 102 \\ \hline \end{array}$$

If you have trouble finding the amount you rounded-up, check the section of rounding up in this chapter where we saw how the last digits add up to 10 and the remaining digits add up to 9.

Step 2 - Now subtract from left to right the amount you rounded-up to.

So if you subtract 4000 from 4530, you get 530. Rounding up made the subtraction process easy.

$$\begin{array}{r} 4530 \\ -4000 + 102 \\ \hline 530 \end{array}$$

Step 3 - Now add the amount you rounded-up with the subtracted amount.

So add 102 to 530. Add from left to right.

$$\begin{array}{r} 530 \\ + 102 \\ \hline 632 \end{array}$$

You get 632 which is your final answer.

Note that we deep dived into the steps so that you have clarity and understanding. When you do the steps in your mind, it should take less than 5 seconds to do this entire calculation.

Example 2

Let us try another example. Subtract 4998 from 7520.

$$\begin{array}{r} 7520 \\ -4998 \\ \hline ----- \end{array}$$

Again doing LR subtraction without rounding up would mean you have to borrow so many numbers as you calculate. So rounding up will simplify the mental process.

Step 1 - Round up the second number.

$$7520 - 4998 = -(5000 - 2)$$

So when you round up 4998, you have 5000, and the amount you rounded-up is 2.

$$7520 - 5000 + 2$$

Step 2 - Subtract from left to right the amount you rounded-up to.

So if you subtract 5000 from 7520, you get 2520. Rounding up made the subtraction process easy.

$$7520 - 5000 + 2 = 2520$$

Step 3 - Now add the amount you rounded-up with the subtracted amount.

So adding 2 to 2520 we get 2522.

$$2520 + 2 = 2522$$

You get 2522 which is your final answer.

Example 3

Let us try another example. Subtract 2796 from 8734.

Remember to round up the second number and then do the subtraction.

Take a second to apply the technique by yourself as fast as you can. Once you have the answer, you can check the steps below to see if you got your answer right.

Step 1 - Round up the second number.

$$\begin{array}{r} 8734 \\ -2796 \end{array} = -(3000 - 204)$$

So when you round up 2796, you have 3,000, and the amount you rounded-up is 204.

$$\begin{array}{r} 8734 \\ -3000 + 204 \end{array}$$

If you have trouble finding the amount you rounded-up, check the section of rounding up in this chapter where we saw how the last digits add up to 10 and the remaining digits add up to 9.

Step 2 - Now subtract from left to right the amount you rounded-up to.

So if you subtract 3000 from 8734, you get 5734. Rounding up made the subtraction process easy.

```
    8 7 3 4
  - 3 0 0 0  + 204
    5 7 3 4
```

Step 3 - Add the amount you rounded-up with the subtracted amount.

So add 204 to 5734. Add from left to right.

```
    5 7 3 4
  +   2 0 4
    5 9 3 8
```

You get 5938 which is your final answer.

If you got your answer wrong, don't worry. Just revisit the techniques and examples we covered in this chapter.

If you have any questions, you can ask it in the math Q & A section of the community by going **ofpad.com/mathqa** and we will make sure to respond. If you have not become a member of the community yet, watch the video **ofpad.com/communityguide** to get started.

When Should You Round Up For Subtraction?

Just like addition, should you always be rounding up before subtracting or should you do it only during specific situations? Sometimes rounding up simplifies the subtraction, and at other times rounding up just complicates the subtraction and creates an unnecessary step. So how do you decide when to round up and when not to round up?

The short answer is that **rounding up should only be done when you have to borrow** a lot of numbers for subtraction just like how you had to carry over a lot of numbers for addition.

Rounding up is more valuable for subtraction than it is for addition because borrowing numbers in your head during subtraction is a lot harder to do compared to carrying over numbers during addition.

In subtraction you round up the number when you have to borrow a lot of numbers. Let us take for example 53,441 - 49,898.

$$\begin{array}{r} 5\,3\,4\,4\,1 \\ -\,4\,9\,8\,9\,8 \\ \hline \end{array}$$

In this subtraction, except for the first numbers (5 - 4) you have to borrow a number during every step. So it makes sense to round up the second number before doing the subtraction.

Rounding up the second number 49,898 by 102 we get 50,000.

$$\begin{array}{r} 5\,3\,4\,4\,1 \\ -\,4\,9\,8\,9\,8 \\ \hline \end{array} = -(50{,}000 - 102)$$

Subtracting 53,441 - 50,000 we get 3441.

$$\begin{array}{r} 5\,3\,4\,4\,1 \\ -\,5\,0\,0\,0\,0 \;+\; 102 \\ \hline 3\,4\,4\,1 \end{array}$$

Adding 102 we get 3,543.

$$\begin{array}{r}3441\\+102\\\hline 3543\end{array}$$

Now try doing the same subtraction 53,441 - 49,898 without rounding up. You might find it more strenuous to do the subtraction without round up.

Do you round up before doing every subtraction? Not really.

Subtract 9889 - 4343.

$$\begin{array}{r}9889\\-4343\\\hline \end{array}$$

You can do the straight forward LR subtraction here to get the answer. You can round up and solve the math but it will only result in an unnecessary step.

So remember to only round up when you have to borrow numbers. If you don't have to borrow any number you will be better off doing the subtraction without rounding up.

Subtraction Exercises

Download the rich PDFs for these exercises from **ofpad.com/mathexercises**.

01) 77
   - 55

02) 81
   - 61

03) 54
   - 25

04) 758 - 482	09) 905 - 227	14) 9880 - 2941
05) 740 - 424	10) 9125 - 5305	15) 8827 - 8718
06) 919 - 872	11) 5487 - 1120	16) 5508 - 2741
07) 867 - 798	12) 8017 - 7676	17) 6991 - 1399
08) 709 - 202	13) 8450 - 8109	18) 7610 - 3171

Subtraction Answers

01) 22
02) 20
03) 29
04) 276
05) 316
06) 47
07) 69
08) 507
09) 678
10) 3,820
11) 4,367
12) 341
13) 341
14) 6,939
15) 109
16) 2,767
17) 5,592
18) 4,439

Chapter 9 – Remembering Numbers

When you do mental math, you will often be faced with a problem of remembering numbers.

For example, if you are adding 45,982 + 44,863, you have to find a way to hold both numbers in your head. You have to remember 10 digits and that is just the numbers you are trying to add.

Apart from this you also need to remember the digits of your answer, which increases the numbers you must hold in your memory to a total of 15 digits.

If you are applying the DS or DD method to verify your answer, you need to continue to hold these 15 digits in your head.

To apply the DS or DD method, you must remember an additional 3 to 5 digits. This increases the maximum number of digits you must remember at any given time to a whopping 20 digits.

Doing mental math is extremely easy but holding these numbers in your memory is a whole different story.

I have created a method called "The Math Palace Technique" that will make remembering numbers effortless. Its adaption of the memory palace method that has existed for centuries, but we will be using it to remember numbers as we calculate.

The technique is still very new. So before I release it to a wider audience, I am collecting feedback from others who have applied it. I am giving "The Math Palace Technique" away in its current form to everyone who leaves a review for the Mental math book.

Once you have left the review you can get The Math Palace Technique from this link: https://ofpad.com/math-palace

Download Link: https://ofpad.com/math-palace

Chapter 10 - LR Multiplication

In this chapter, we will look at how to multiply any set of numbers fast using the LR Method. You have already seen the power of calculating from left to right in the previous chapters for addition and subtraction. In this chapter, we will apply the same concept to multiplication.

The secret of mental multiplication like addition and subtraction is to multiply from left to right instead of right to left. The beauty of this method is that it is a generic technique that can be applied to any pairs of numbers. There is one prerequisite for this chapter. You need to know your multiplication tables.

Below is a quick refresher if you need it:

X	1	2	3	4	5	6	7	8	9
1	1	2	3	4	5	6	7	8	9
2	2	4	6	8	10	12	14	16	18
3	3	6	9	12	15	18	21	24	27
4	4	8	12	16	20	24	28	32	36
5	5	10	15	20	25	30	35	40	45
6	6	12	18	24	30	36	42	48	54
7	7	14	21	28	35	42	49	56	63
8	8	16	24	32	40	48	56	64	72
9	9	18	27	36	45	54	63	72	81

When you look at the examples in this chapter and later do the practice, learn to visualise the numbers in your mind.

That is where practice comes in. The more you practice, the better memory habits you will develop, which will make remembering numbers easier.

One-digit Multiplier

We will start off by applying LR Multiplication to one-digit multipliers.

Example 1

Let us try an example. Multiply 5321 by 4.

$$\begin{array}{r} 5\,3\,2\,1 \\ \times 4 \\ \hline \_\,\_\,\_\,\_\,\_ \end{array}$$

Multiply from left to right. Multiplying 5 x 4 gives us 20.

$$\begin{array}{r} 5\,3\,2\,1 \\ \times 4 \\ \hline 2\,0\,\_\,\_\,\_ \end{array}$$

Multiplying 3 x 4 gives us 12.

$$\begin{array}{r} 5\,3\,2\,1 \\ \times 4 \\ \hline 2\,1\,2\,\_\,\_ \end{array}$$

Multiplying 2 x 4 gives us 8.

```
   5 3 2 1
 x       4
 ─────────
   2 1 2 8 _
```

Multiplying 1 x 4 gives us 4.

```
   5 3 2 1
 x       4
 ─────────
   2 1 2 8 4
```

Now you have your final answer 21,284.

Example 2

Let us try another example. Multiply 7142 by 6.

```
   7 1 4 2
 x       6
 ─────────
   _ _ _ _ _
```

Multiply from left to right.

When doing this calculation, recall how we covered carrying numbers over when we multiplied by 11. The same applies here and in other techniques in this book.

Multiplying 7 x 6 gives us 42.

```
   7 1 4 2
 x       6
 ─────────
   4 2 _ _ _
```

Multiplying 1 x 6 gives us 6.

```
      7 1 4 2
  x         6
  ─────────────
      4 2 6 _ _
```

Multiplying 4 x 6 gives us 24.

```
      7 1 4 2
  x         6
  ─────────────
      4 2 6 _ _
            2 4
```

Carry over the 2, so the 6 becomes 8.

```
      7 1 4 2
  x         6
  ─────────────
      4 2 8 4 _
```

Multiply 2 x 6 gives us 12.

```
      7 1 4 2
  x         6
  ─────────────
      4 2 8 4 _
              1 2
```

Carry over the 1, so the 4 becomes 5.

```
      7 1 4 2
  x         6
  ─────────────
      4 2 8 5 2
```

And you have the final answer 42,852.

When you have to carry over numbers, the difficult part may be holding the numbers you already calculated in your memory. Saying the numbers out loud will help free up some of the memory. However, with practice, you will be able to effortlessly remember even without doing this, because practice will improve your concentration and memory. When you are doing the examples in this book, you might be tempted to glance at the problem to remind yourself of where you are at. This is fine but try to hold the numbers in your head. It will become easier the more you practice doing it.

Example 3

Let us try another example.

Multiply 9432 by 3.

$$\begin{array}{r} 9432 \\ \times 3 \\ \hline \_\_\_\_\_ \end{array}$$

Multiply from left to right.

Take a second to apply the technique by yourself as fast as you can. Once you have the answer, you can check the steps below to see if you got your answer right.

Multiplying 9 x 3 gives us 27.

$$\begin{array}{r} 9432 \\ \times 3 \\ \hline 27\_\_\_ \end{array}$$

Multiplying 4 x 3 gives us 12.

```
   9 4 3 2
x        3
──────────
   2 7 _ _ _
   1 2
```

Carry over the 1, so the 7 becomes 8.

```
   9 4 3 2
x        3
──────────
   2 8 2 _ _
```

Multiplying 3 x 3 gives us 9.

```
   9 4 3 2
x        3
──────────
   2 8 2 9 _
```

Multiplying 2 x 3 gives us 6.

```
   9 4 3 2
x        3
──────────
   2 8 2 9 6
```

The final answer is 28,296.

If you got your answer wrong, don't worry. Just revisit the techniques and examples we covered in this chapter.

If you have any questions, you can ask it in the math Q & A section of the community by going **ofpad.com/mathqa**. We will make sure to respond. If you have not become a member of the community yet, watch the video **ofpad.com/communityguide** to get started.

At first, mental math might seem like a thing only a genius can do. However, it is like learning how to ride a bicycle - you cannot forget it, once you have learnt it.

Exercises

Download the rich PDFs for these exercises from **ofpad.com/mathexercises**.

01) 86 x 5	07) 168 x 5	13) 8013 x 6
02) 45 x 7	08) 906 x 4	14) 5088 x 9
03) 60 x 9	09) 520 x 7	15) 7895 x 9
04) 510 x 7	10) 1816 x 4	16) 2766 x 3
05) 398 x 6	11) 3619 x 7	17) 5781 x 5
06) 645 x 9	12) 8824 x 5	18) 5451 x 7

Answers

01) 430
02) 315
03) 540
04) 3,570
05) 2,388
06) 5,805

07) 840
08) 3,624
09) 3,640
10) 7,264
11) 25,333
12) 44,120

13) 48,078
14) 45,792
15) 71,055
16) 8,298
17) 28,905
18) 38,157

One-digit Multiplier With Rounding Up

Just like in LR addition and subtraction, rounding up and multiplying is especially useful when the numbers end in 7, 8 or 9.

Method

The method is simple:

Step 1 - Round up the number.

Step 2 - Multiply from left to right.

Step 3 - Multiply the amount you rounded-up.

Step 4 - Subtract the numbers from the previous two steps.

Let us look at an example to illustrate the method.

Example 1

Multiply 68 by 3.

$$\begin{array}{r} 6\,8 \\ \times\,3 \\ \hline - - - \end{array}$$

Step 1 - Round up the number.

$$\begin{array}{r} 6\,8\ (70-2) \\ \times\,3 \\ \hline - - - \end{array}$$

So round up 68 to 70. You have rounded-up by 2.

```
  70 – 2
   x 3
  ─────
  – – –
```

Step 2 - Multiply from left to right.

Multiply 70 by 3, and you have 210.

```
  70 – 2
   x 3
  ─────
  2 1 0
```

Step 3 - Multiply the amount you rounded-up.

Multiplying 2 x 3 gives us 6.

```
  70 –  2
   x 3  x 3
  ───── ───
  2 1 0  6
```

Step 4 - Subtract the numbers from the previous two steps.

Subtracting 210 – 6 we get 204.

```
  70    2
   x 3  x 3
  ───── ───
  2 1 0 – 6 = 204
```

And now you have the answer as 204.

```
  6 8
   x 3
  ─────
  2 0 4
```

The entire multiplication process has been greatly simplified because you rounded-up.

Example 2

Let us try another example.

Multiply 96 by 7.

$$\begin{array}{r} 96 \\ \times\ 7 \\ \hline \end{array}$$

Step 1 - Round up the number.

$$\begin{array}{r} 96 = 100 - 4 \\ \times\ 7 \qquad\qquad \\ \hline \end{array}$$

So we round up 96 to 100. We have rounded-up by 4.

Step 2 - Multiply from left to right.

$$\begin{array}{r} 100 - 4 \\ \times\ 7 \qquad \\ \hline \end{array}$$

Multiplying 100 by 7 gives us 700.

$$\begin{array}{r} 100 - 4 \\ \times\ 7 \qquad \\ \hline 700 \qquad \end{array}$$

Step 3 - Multiply the amount you rounded-up.

Multiplying 4 x 7 gives us 28.

```
  1 0 0  −   4
  x   7     x 7
  ─────     ───
  7 0 0     2 8
```

Step 4 - Subtract the numbers from the previous two steps.

Subtracting 700 − 28 we get 672.

```
  1 0 0       4
  x   7     x 7
  ─────     ───
  7 0 0  −  2 8  = 672
```

672 is the result of the original multiplication.

```
    9 6
    x 7
    ───
    6 7 2
```

If you have trouble in this last step of subtraction, remember the rule for rounding up we saw in the previous chapter. The same rule can be applied to do the subtraction quickly. Notice the last digits (8 in 28 and 2 in 672) add up to 10 and the other digits (2 in 28 and 7 and 672) add up to 9. The rounding up rule can be applied to subtraction of numbers like this, and you can find out the answer without explicitly doing the subtraction.

A seemingly complicated multiplication problem has become so simple because we just rounded-up.

Example 3

Let us try another example.

Multiply 398 by 9.

$$\begin{array}{r} 3\,9\,8 \\ \times\quad 9 \\ \hline \_\_\_\_ \end{array}$$

Take a second to apply the technique by yourself as fast as you can. Once you have the answer, you can check the steps below to see if you got your answer right.

Step 1 - Round up the number.

$$\begin{array}{r} 3\,9\,8 = (400-2) \\ \times\quad 9 \\ \hline \_\_\_\_ \end{array}$$

So we round up 398 to 400. We have rounded-up by 2.

$$\begin{array}{r} 4\,0\,0\,-\,2 \\ \times\quad 9 \\ \hline \_\_\_\_ \end{array}$$

Step 2 - Multiply from left to right.

Multiply 400 by 9, and you have 3600.

$$\begin{array}{r} 4\,0\,0\,-\,2 \\ \times\quad 9 \\ \hline 3\,6\,0\,0 \end{array}$$

Step 3 - Multiply the amount you rounded-up.

Multiplying 2 x 9 gives us 18.

```
  400    – 2
x   9     x 9
─────    ────
 3600     18
```

Step 4 - Subtract the numbers from the previous two steps.

Subtract 3600 – 18 to get 3582.

```
  400      2
x   9    x 9
─────   ────
 3600 – 18  = 3582
```

3582 is answer of the original multiplication.

```
   398
 x   9
 ─────
  3582
```

If you have trouble in this last step of subtraction, remember the rule we saw in the previous chapter where the last digits (8 in 18 and 2 in 3592) add up to 10 and the other digits (1 in 18 and 8 and 3582) add up to 9. It will greatly simplify the last step, and you will be able to find the answer without explicitly doing any subtraction.

If you got your answer wrong, don't worry. Just revisit the techniques and examples we covered in this chapter.

If you have any questions, you can ask it in the math Q & A section of the community by going **ofpad.com/mathqa** and we will make sure to respond. If you have not become a member of the community yet, watch the video **ofpad.com/communityguide** to get started.

Once you finish practicing, move on to the next section.

Exercises

Download the rich PDFs for these exercises from **ofpad.com/mathexercises**.

01) 57
  x 8
	07) 498
	  x 3
		13) 4998
		  x 6

02) 28
  x 3
	08) 298
	  x 8
		14) 1999
		  x 7

03) 67
  x 8
	09) 197
	  x 7
		15) 8999
		  x 3

04) 798
  x   3
	10) 4997
	  x 4
		16) 6997
		  x 9

05) 798
  x 8
	11) 6998
	  x 5
		17) 4997
		  x 6

06) 298
  x   9
	12) 2998
	  x 4
		18) 6998
		  x 8

Answers

01) 456	05) 6,384	09) 1,379
02) 84	06) 2,682	10) 19,988
03) 536	07) 1,494	11) 34,990
04) 2,394	08) 2,384	12) 11,992

13) 29,988   15) 26,997   17) 29,982
14) 13,993   16) 62,973   18) 55,984

LR Multiplication – Two Digit Multiplier

We looked at multiplication using a 1 digit multiplier. Let us look at a 2 digit multiplier now. To multiply by a 2 digit multiplier.

Method

Step 1 - Break the multiplier.

Step 2 - Multiply from left to right.

Step 3 - Add the individual answers together.

Let us look at an example.

Example 1

So let us multiply 36 by 32.

$$\begin{array}{r} 3\,6 \\ \times\,3\,2 \\ \hline \_\,\_\,\_\,\_ \end{array}$$

Step 1 - Break the multiplier.

32 is broken down as 30 + 2.

$$\begin{array}{r} 3\,6 \\ \times\,3\,2 = (30 + 2) \\ \hline \_\,\_\,\_\,\_ \end{array}$$

Step 2 - Multiply from left to right.

Multiplying 36 x 30 gives us 1080.

$$\begin{array}{r} 36 \\ \times 30 + 2 \\ \hline 1080 \end{array}$$

Multiplying 36 x 2 gives us 72.

$$\begin{array}{rr} 36 & 36 \\ \times 30 + & \times 2 \\ \hline 1080 & 72 \end{array}$$

Step 3 - Add the individual answers together.

Adding 1080 + 72 gives us 1152.

$$\begin{array}{rr} 36 & 36 \\ \times 30 & \times 2 \\ \hline 1080 + & 72 = 1152 \end{array}$$

We have the final answer 1152.

$$\begin{array}{r} 36 \\ \times 32 \\ \hline 1152 \end{array}$$

Example 2

Let us try another example. Multiply 26 by 23.

Mental Math: Tricks to Become a Human Calculator

$$\begin{array}{r} 26 \\ \times\,23 \\ \hline --- \end{array}$$

Step 1 - Break the multiplier.

$$\begin{array}{r} 26 \\ \times\,23 \\ \hline --- \end{array} = (20 + 3)$$

So 23 is broken down as 20 + 3.

$$\begin{array}{r} 26 \\ \times\,20\,+\,3 \\ \hline --- \end{array}$$

Step 2 - Multiply from left to right.

Multiplying 26 x 20 gives us 520.

$$\begin{array}{r} 26 \\ \times\,20\,+\,3 \\ \hline 520 \end{array}$$

Multiplying 26 x 3 gives us 78.

$$\begin{array}{rr} 26 & 26 \\ \times\,20\,+ & \times\,3 \\ \hline 520 & 78 \end{array}$$

Step 3 - Add the individual answers together.

Adding 520 + 78 gives us 598.

$$\begin{array}{rr} 26 & 26 \\ \times 20 & \times 3 \\ \hline 520 + & 78 = 598 \end{array}$$

We have the final answer 598.

$$\begin{array}{r} 26 \\ \times 23 \\ \hline 598 \end{array}$$

Example 3

Let us try another example.

Let us multiply 72 by 41.

$$\begin{array}{r} 72 \\ \times 41 \\ \hline \end{array}$$

Take a second to apply the technique by yourself as fast as you can. Once you have the answer, you can check the steps below to see if you got your answer right.

Step 1 - Break the multiplier.

$$\begin{array}{r} 72 \\ \times 41 \\ \hline \end{array} = (40 + 1)$$

41 is broken down as 40 + 1.

$$\begin{array}{r}72\\ \times\,40+1\\ \hline ----\end{array}$$

Step 2 - Multiply from left to right.

Multiplying 72 x 40 gives us 2880.

$$\begin{array}{r}72\\ \times\,40+1\\ \hline 2880\end{array}$$

Multiplying 72 x 1 gives us 72.

$$\begin{array}{rr}72 & 72\\ \times\,40+ & \times\,1\\ \hline 2880 & 72\end{array}$$

Step 3 - Add the individual answers together.

Adding 2880 + 72 gives us 2952.

$$\begin{array}{rr}72 & 72\\ \times\,40 & \times\,1\\ \hline 2880\,+ & 72 = 2952\end{array}$$

You have the final answer 2952.

$$\begin{array}{r}72\\ \times\,41\\ \hline 2952\end{array}$$

If you got your answer wrong, don't worry. Just revisit the techniques and examples we covered in this chapter.

If you have any questions, you can ask it in the math Q & A section of the community by going **ofpad.com/mathqa**. If you have not become a member of the community yet, watch the video **ofpad.com/communityguide** to get started.

Once you finish practicing, move on to the next section.

Exercises

Download the rich PDFs for these exercises from **ofpad.com/mathexercises**.

01) 43
  x 11
	06) 36
	  x 24
		11) 56
		  x 19

02) 36
  x 71
	07) 42
	  x 36
		12) 13
		  x 98

03) 89
  x 14
	08) 60
	  x 22
		13) 59
		  x 98

04) 29
  x 48
	09) 72
	  x 47
		14) 50
		  x 90

05) 32
  x 24
	10) 19
	  x 91
		15) 99
		  x 13

16) 89	17) 13	18) 53
x 25	x 77	x 24

Answers

01) 473	07) 1,512	13) 5,782
02) 2,556	08) 1,320	14) 4,500
03) 1,246	09) 3,384	15) 1,287
04) 1,392	10) 1,729	16) 2,225
05) 768	11) 1,064	17) 1,001
06) 864	12) 1,274	18) 1,272

LR Multiplication – Two Digit Multiplier With Round Up

Rounding up and multiplying is also useful for two-digit multipliers when the numbers end in 7, 8 or 9.

Method

To do that:

Step 1 - Round up a number.

Step 2 - Multiply the rounded-up value and the amount you rounded-up from left to right.

Step 3 - Subtract the two numbers.

Let us look at an example.

Example 1

Multiply 87 by 99.

$$\begin{array}{r}87\\ \times\ 99\\ \hline ----\end{array}$$

Step 1 - Round up a number.

$$\begin{array}{r}87\\ \times\ 99\\ \hline ----\end{array} = (100-1)$$

So we round up 99 to 100. We rounded-up by 1.

$$\begin{array}{r}87\\ \times 100\\ \hline ----\end{array} - 1$$

Step 2 - Multiply the rounded-up value and the amount you rounded-up from left to right.

So multiply 87 with 100 to get 8700.

$$\begin{array}{r}87\\ \times 100\\ \hline 8700\end{array} - 1$$

Multiply 87 with 1 to get 87.

$$\begin{array}{r}87\\ \times 100\\ \hline 8700\end{array} \quad \begin{array}{r}87\\ \times 1\\ \hline 87\end{array}$$

Step 3 - Subtract the two numbers.

Subtract 8700 − 87 to get 8613.

$$\begin{array}{rr} 87 & 87 \\ \times\,100 & \times\,1 \\ \hline 8700 & -\;87 \end{array} = 8613$$

If you have trouble with this subtraction step remember that the last digits (7 in 87 and 3 in 8613) add to give 10 and the other digits (8 in 87 and 1 in 8613) adds up to give 9.

$$\begin{array}{r} 87 \\ \times\;99 \\ \hline 8613 \end{array}$$

Your final answer is 8613.

Example 2

Let us look at another example.

Multiply 41 by 57.

$$\begin{array}{r} 41 \\ \times\;57 \\ \hline \;\;\;-\,-\,-\,- \end{array}$$

Step 1 - Round up a number.

$$\begin{array}{r} 41 \\ \times\;57 \end{array} = (60-3)$$

$$-\,-\,-\,-$$

So we round up 57 to 60. We rounded-up by 3.

$$\begin{array}{r} 41 \\ \times\ 60 - 3 \\ \hline ----\end{array}$$

Step 2 - Multiply the rounded-up value and the amount you rounded-up from left to right.

So multiply 41 with 60 to get 2460.

$$\begin{array}{r} 41 \\ \times\ 60 - 3 \\ \hline 2460 \end{array}$$

Then multiply 41 with 3 to give 123.

$$\begin{array}{rr} 41 & 41 \\ \times\ 60 - & \times\ 3 \\ \hline 2460 & 123 \end{array}$$

Step 3 - Subtract the two numbers.

Subtracting 2460 – 123 gives us 2337.

$$\begin{array}{rr} 41 & 41 \\ \times\ 60 & \times\ 3 \\ \hline 2460 - 123 & = 2337 \end{array}$$

Your final answer is 2337.

$$\begin{array}{r} 41 \\ \times\ 57 \\ \hline 2337 \end{array}$$

Example 3

Let us look at another example.

Multiply 67 by 38.

$$\begin{array}{r} 67 \\ \times\ 38 \\ \hline \end{array}$$

Take a second to apply the technique by yourself as fast as you can. Once you have the answer, you can check the steps below to see if you got your answer right.

Step 1 - Round up a number.

$$\begin{array}{r} 67 \\ \times\ \underline{38} \\ \hline \end{array} = (40-2)$$

So we round up 38 to 40. We rounded-up by 2.

$$\begin{array}{r} 67 \\ \times\ \underline{40} - 2 \\ \hline \end{array}$$

Step 2 - Multiply the rounded-up value and the amount you rounded-up from left to right.

Multiplying 67 by 40 gives us 2680.

$$\begin{array}{r} 67 \\ \times\ 40 - 2 \\ \hline 2680 \end{array}$$

Next multiplying 67 by 2 gives us 134.

$$\begin{array}{rr} 67 & 67 \\ \times\ 40 - & \times 2 \\ \hline 2680 & 134 \end{array}$$

Step 3 - Subtract the two numbers.

Subtracting 2680 − 134 gives us 2546.

$$\begin{array}{rr} 67 & 67 \\ \times\ 40 & \times 2 \\ \hline 2680 & -134 \end{array} = 2546$$

Now you have 2546 which is your final answer.

$$\begin{array}{r} 67 \\ \times\ 38 \\ \hline 2546 \end{array}$$

If you got your answer wrong, don't worry. Just revisit the techniques and examples we covered in this chapter.

If you have any questions, you can ask it in the math Q & A section of the community by going **ofpad.com/mathqa** and we will make sure to respond. If you have not become a member of the community yet, watch the video **ofpad.com/communityguide** to get started.

Once you finish practicing, move on to the next section.

Exercises

Download the rich PDFs for these exercises from **ofpad.com/mathexercises**.

01) 67 x 76	x 38	x 38
02) 38 x 82	08) 58 x 83	14) 18 x 96
03) 29 x 29	09) 18 x 90	15) 37 x 65
04) 67 x 61	10) 69 x 36	16) 49 x 22
05) 19 x 91	11) 67 x 42	17) 79 x 17
06) 49 x 54	12) 79 x 84	18) 49 x 34
07) 38	13) 58	

Answers

01) 5,092	07) 1,444	13) 2,204
02) 3,116	08) 4,814	14) 1,728
03) 841	09) 1,620	15) 2,405
04) 4,087	10) 2,484	16) 1,078
05) 1,729	11) 2,814	17) 1,343
06) 2,646	12) 6,636	18) 1,666

LR Multiplication After Factoring

Like rounding up, another technique to use before you apply the LR method is to factor the number before multiplying it. Factoring a number means breaking it down into one-digit numbers, which when later multiplied together will give the original number.

For example, 48 can be factored as 6 x 8 or 12 x 4. However, we will prefer to use single digit factors as the intention of factoring the numbers is to simplify the multiplication.

Method

To apply LR method with factoring:

Step 1 - Factor one of the numbers.

Step 2 - Multiply the number with the first factor (left to right).

Step 3 - Multiply the product with the second factor (left to right).

Let us look at an example.

Example 1

Multiply 45 by 22.

Mental Math: Tricks to Become a Human Calculator

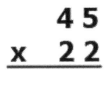

Step 1 - Factor one of the numbers.

Let us factor 22. So, 22 can be factored as 11 x 2.

$$\begin{array}{r} 45 \\ \times\ 11 \times 2 \\ \hline --- \end{array}$$

Step 2 - Multiply the number with the first factor (left to right).

So, multiply 45 by 11. You get 495. Since we are multiplying by 11, you can apply the rule to multiply by 11 instead of doing an LR multiplication.

$$\begin{array}{r} 45 \\ \times\ 11 \times 2 \\ \hline 495 \end{array}$$

Step 3 - Multiply the product with the second factor (left to right).

So now take the 495 and multiply it by 2. Multiply from left to right.

So 495 multiplied by 2 gives us 990.

$$\begin{array}{r} 495 \\ \times 2 \\ \hline 990 \end{array}$$

990 is your final answer.

You could have factored 45 as 9 x 5 and multiplied it by 22 instead of factoring 22. Also, you could have used 2 as the first factor and 11 as the second factor. The choice is always up to you, but the easier and smaller the numbers, the faster you will calculate.

Example 2

Let us look at another example.

Multiply 21 by 63.

$$\begin{array}{r} 21 \\ \times 63 \\ \hline - - - \end{array}$$

Step 1 - Factor one of the numbers.

Let us factor 63. So 63 can be factored as 7 x 9.

```
    2 1
x   7 x 9
---------
  _ _ _
```

Step 2 - Multiply the number with the first factor (left to right).

So multiply 21 by 7 left to right. You get 147.

```
    2 1
x   7 x 9
---------
   1 4 7
```

Step 3 - Multiply the product with the second factor (left to right).

So now take the 147 and multiply it by 9. Multiply from left to right.

So 147 multiplied by 9 gives us 1323.

```
   1 4 7
x      9
--------
  1 3 2 3
```

Your final answer is 1323.

You could have factored 21 into 7 x 3 instead of factoring 63 into 7 x 9 and solved this problem, and some would have found that easier.

Example 3

Let us look at another example.

Multiply 42 by 36.

$$\begin{array}{r} 42 \\ \times\ 36 \\ \hline --- \end{array}$$

Take a second to apply the technique by yourself as fast as you can. Once you have the answer, you can check the steps below to see if you got your answer right.

Step 1 - Factor one of the numbers.

Let us factor 36. So 36 can be factored as 6 x 6.

$$\begin{array}{r} 42 \\ \times\ \ \ 6 \times 6 \\ \hline --- \end{array}$$

Step 2 - Multiply the number with the first factor (left to right).

So multiply 42 by 6 left to right. You get 252.

$$\begin{array}{r} 42 \\ \times\ \ \ 6 \times 6 \\ \hline 252 \end{array}$$

Step 3 - Multiply the product with the second factor (left to right).

So now take the 252 and multiply it by 6. Multiply from left to right. So 252 multiplied by 6 gives us 1512.

$$\begin{array}{r} 252 \\ \times 6 \\ \hline 1512 \end{array}$$

And 1512 is your final answer.

You could have factored 42 into 7 x 6 instead of factoring 36 into 6 x 6 and solved this problem, and you would have got the same answer.

If you got your answer wrong, don't worry. Just revisit the techniques and examples we covered in this chapter.

Read this chapter again if necessary. Then go to the practice section and complete the exercises.

You might have understood the technique, but only practice will make using these techniques effortless and easy.

If you have any questions, you can ask it in the math Q & A section of the community by going **ofpad.com/mathqa** and we will make sure to respond. If you have not become a member of the community yet, watch the video **ofpad.com/communityguide** to get started.

If you have enjoyed the book so far, do leave a review on Amazon by visiting **ofpad.com/mathbook**.

If you did not enjoy the book so far, and if you have any general suggestions to improve the book or specific feedback for this chapter, do let us know at **ofpad.com/feedback**. If your feedback helps us improve the book even in small ways, we will thank you by sending a free review copy of our next product when it becomes available.

Once you finish practicing, move on to the next section.

Exercises

Download the rich PDFs for these exercises from **ofpad.com/mathexercises**.

01) 38 x 77	07) 28 x 36	13) 79 x 72
02) 37 x 63	08) 68 x 33	14) 38 x 56
03) 37 x 54	09) 58 x 16	15) 87 x 24
04) 89 x 54	10) 27 x 81	16) 17 x 42
05) 49 x 99	11) 77 x 24	17) 67 x 64
06) 89 x 56	12) 77 x 49	18) 17 x 18

Answers

01) 2,926
02) 2,331
03) 1,998
04) 4,806
05) 4,851
06) 4,984
07) 1,008
08) 2,244
09) 928
10) 2,187
11) 1,848
12) 3,773
13) 5,688
14) 2,128
15) 2,088
16) 714
17) 4,288
18) 306

Chapter 11 - Stem Method

In this chapter we will look at how to do fast multiplication using the stem method. It is called the stem method because we will use a stem number to do the multiplication.

Stem Multiplication

To multiply using the stem number:

Step 1 - Decide a stem number.

Step 2 - Find the difference of the stem number from the multiplicand and the multiplier.

Step 3 - Add the multiplicand with the difference of the multiplier **OR** add the multiplier with the difference of the multiplicand. (**Note:** Both differences will give the same result).

Step 4 - Multiply the stem number with the result of the addition.

Step 5 - Multiply the difference of the stem number from the multiplicand with the difference of the stem number from the multiplier.

Step 6 - Add the products of the last two steps to get your final answer.

We will start off with a few easy examples and see how to multiply numbers up to 10.

One-by-One Multiplication

Example 1

Let's try multiplying 9 and 7.

$$9 \times 7$$

Step 1 – Decide a stem number.

We will use 10 as the stem number.

Stem 10
$$9 \times 7$$

Step 2 - Find the difference of the stem number from the multiplicand and the multiplier.

For the multiplicand 9 the difference of 9 – 10 gives us -1.

Stem 10
$$\begin{array}{ccc} 9 & \times & 7 \\ -1 & & \end{array}$$

For the multiplier 7 the difference of 7 – 10 gives us -3.

Stem 10
$$\begin{array}{ccc} 9 & \times & 7 \\ -1 & & -3 \end{array}$$

Step 3 - Add the multiplicand with the difference of the multiplier **OR** add the multiplier with the difference of the multiplicand.

We will be adding diagonally.

Let us choose to add 9 with -3 to get 6.

$$\begin{array}{c} \text{Stem 10} \\ 9 \quad \times \quad 7 \;=\; 6 \\ -1 \quad\quad\; -3 \end{array}$$

Had you chosen to add -1 with 7 you would have got the same 6.

$$\begin{array}{c} \text{Stem 10} \\ 9 \quad \times \quad 7 \;=\; 6 \\ -1 \quad\quad\; -3 \end{array}$$

Since both diagonals give the same result, pick the easier of the two additions.

Step 4 - Multiply the stem number with the result of the addition.

So if we multiply the stem number 10 with the result of the addition 6 we get 10 x 6 = 60.

$$\begin{array}{c} \text{Stem 10} \\ 9 \quad \times \quad 7 \;=\; 60 \\ -1 \quad\quad\; -3 \end{array}$$

Step 5 - Multiply the difference of the stem number from the multiplicand with the difference of the stem number from the multiplier.

So we multiply -1 x -3 to get +3.

Stem 10

$$9 \times 7 = 60$$
$$-1 \times -3 = +3$$

Step 6 - Add the products of the last two steps to get your final answer.

So adding 60 with 3 we get 63.

Stem 10

$$9 \times 7 = 60$$
$$-1 \times -3 = \underline{+\ 3}$$
$$63$$

63 is our final answer.

$$9 \times 7 = 63$$

Let us try another example.

Example 2

Let's try multiplying 4 and 6.

$$4 \times 6$$

Step 1 – Decide a stem number.

If we use 10 as the stem number, we will have a problem.

Stem 10
$$4 \times 6$$

In step 2, we will find the difference of the stem number from the multiplicand and the multiplier.

For the multiplicand 4, the difference of 4 − 10 gives us -6.

Stem 10
4 x 6
-6

For the multiplier 6, the difference of 6 − 10 gives us -4.

Stem 10
4 x 6
-6   -4

This is the same as the original problem. Because the numbers are so far away from 10, using the 10 as the stem number does not simplify the problem in anyway.

So instead of 10 let us use the number 5 as the stem number. Unlike 10, if we use 5, it is a lot closer to both 6 and 4.

Stem 5
4 x 6

In step 2, we will find the difference of the stem number from the multiplicand and the multiplier.

For the multiplicand 4 the difference of 4 − 5 gives us -1.

For the multiplier 6 the difference of 6 − 5 gives us +1.

Stem 5
4 x 6
-1   +1

Step 3 - Add the multiplicand with the difference of the multiplier **OR** add the multiplier with the difference of the multiplicand.

We will be adding diagonally.

Let us choose to add 4 with +1 to get +5.

$$\begin{array}{c} \textbf{Stem 5} \\ 4 \quad \times \quad 6 \quad = 5 \\ -1 \qquad +1 \end{array}$$

Had you chosen to add -1 with 6 you would have got the same +5.

$$\begin{array}{c} \textbf{Stem 5} \\ 4 \quad \times \quad 6 \quad = 5 \\ -1 \qquad +1 \end{array}$$

Step 4 - Multiply the stem number with the result of the addition.

So if we multiply the stem number 5 with the result of the addition 5 we get 5 x 5 = 25.

$$\begin{array}{c} \textbf{Stem 5} \\ 4 \quad \times \quad 6 \quad = 25 \\ -1 \qquad +1 \end{array}$$

Step 5 - Multiply the difference of the stem number from the multiplicand with the difference of the stem number from the multiplier.

So we multiply -1 x +1 to get -1.

Stem 5

$$4 \times 6 = 25$$
$$-1 \times +1 = -1$$

Step 6 - Add the products of the last two steps to get your final answer.

So adding 25 with -1 we get 24.

Stem 5

$$4 \times 6 = 25$$
$$\underline{-1 \times +1 = -1}$$
$$24$$

24 is our final answer.

$$4 \times 6 = 24$$

Let us try another example.

Example 3

Multiply 8 x 9.

$$8 \times 9$$

Take a second to apply the technique by yourself as fast as you can. Once you have the answer, you can check the steps below to see if you got your answer right.

Step 1 – Decide a stem number.

We will use 10 as the stem number.

Stem 10
8 x 9

Step 2 - Find the difference of the stem number from the multiplicand and the multiplier.

For the multiplicand 8 the difference of 8 – 10 gives us -2.

Stem 10
8 x 9
-2

For the multiplier 9 the difference of 9 – 10 gives us -1.

Stem 10
8 x 9
-2 -1

Step 3 - Add the multiplicand with the difference of the multiplier **OR** add the multiplier with the difference of the multiplicand. We will be adding diagonally.

Let us choose to add 8 with -1 to get 7.

Stem 10
8 x 9 = 7
-2 -1

Had you chosen to add -2 with 9 you would have got the same 7.

Stem 10
8 x 9 = 7
-2 -1

Step 4 - Multiply the stem number with the result of the addition.

So if we multiply the stem number 10 with the result of the addition 7 we get 10 x 7 = 70.

$$\begin{array}{c} \textbf{Stem 10} \\ \begin{array}{ccc} 8 & \times & 9 \\ -2 & & -1 \end{array} = 70 \end{array}$$

Step 5 - Multiply the difference of the stem number from the multiplicand with the difference of the stem number from the multiplier.

So we multiply -2 x -1 to get +2.

$$\begin{array}{c} \textbf{Stem 10} \\ \begin{array}{ccc} 8 & \times & 9 & = 70 \\ -2 & \times & -1 & = +2 \end{array} \end{array}$$

Step 6 - Add the products of the last two steps to get your final answer.

So adding 70 with 2 we get 72.

$$\begin{array}{c} \textbf{Stem 10} \\ \begin{array}{ccc} 8 & \times & 9 & = 70 \\ -2 & \times & -1 & = \underline{+2} \\ & & & 72 \end{array} \end{array}$$

72 is our final answer.

$$8 \times 9 = 72$$

Exercises

Download the rich PDFs for these exercises from **ofpad.com/mathexercises**.

01) 5 x 4
02) 6 x 6
03) 9 x 8
04) 6 x 4
05) 9 x 9
06) 7 x 7
07) 5 x 7
08) 7 x 4
09) 6 x 9
10) 3 x 5
11) 11 x 9
12) 8 x 8
13) 11 x 8
14) 8 x 7
15) 4 x 3
16) 6 x 5
17) 9 x 7
18) 8 x 6

Answers

01) 20
02) 36
03) 72
04) 24
05) 81
06) 49
07) 35
08) 28
09) 54
10) 15
11) 99
12) 64
13) 88
14) 56
15) 12
16) 30
17) 63
18) 48

Learn Multiplication Tables?

Now that you have learnt how to multiply numbers up to 10 using the stem method, does this mean that you don't learn your multiplication tables?

The answer is a resounding NO.

You must commit single digit multiplication tables to memory. If you don't, you won't be able to use the other multiplication techniques to do mental math.

The only reason we covered single digit multiplication is because it helps to illustrate stem multiplication.

So far, we covered how to apply this technique for numbers up to 10. Let us see how to apply this technique for bigger numbers.

Two-by-Two Multiplication

Example 1

Multiply 97 x 93.

$$97 \times 93$$

Step 1 – Decide a stem number.

We will use 100 as the stem number because it is a multiple of 10 and it is easier to multiply numbers with 100.

Stem 100
$$97 \times 93$$

Step 2 - Find the difference of the stem number from the multiplicand and the multiplier.

For the multiplicand 97, the difference of 97 – 100 gives us -3.

Stem 100
$$97 \times 93$$
$$-3$$

For the multiplier 93, the difference of 93 – 100 gives us -7.

Stem 100
97 × 93
-3   -7

Step 3 - Add the multiplicand with the difference of the multiplier **OR** add the multiplier with the difference of the multiplicand.

We will be adding diagonally.

Let us choose to add 97 with -7 to get 90.

Stem 100
97 × 93 = 90
-3   -7

Had you chosen to add -3 with 93 you would have got the same 90.

Stem 100
97 × 93 = 90
-3   -7

Since both diagonals give the same result, you can pick the easier of the two.

Step 4 - Multiply the stem number with the result of the addition.

So if we multiply the stem number 100 with the result of the addition 90 we get 9000.

Stem 100

```
97   x   93   = 9000
-3       -7
```

Step 5 - Multiply the difference of the stem number from the multiplicand with the difference of the stem number from the multiplier.

So we multiply -3 x -7 to get +21.

Stem 100

```
97   x   93   = 9000
-3   x   -7   = +  21
```

Step 6 - Add the products of the last two steps to get your final answer.

So adding 9000 with 21 we get 9021.

Stem 100

```
97   x   93   = 9000
-3   x   -7   = +  21
                  9021
```

9021 is our final answer.

$$97 \times 93 = 9021$$

Let us try another example.

Example 2

Multiply 22 x 24

22 x 24

Step 1 – Decide a stem number.

We will use 20 as the stem number, since it is closest to 22 and 24.

Stem 20
22 x 24

Step 2 - Find the difference of the stem number from the multiplicand and the multiplier.

For the multiplicand 22 the difference of 22 – 20 gives us 2.

Stem 20
22 x 24
2

For the multiplier 24 the difference of 24 – 20 gives us 4.

Stem 20
22 x 24
2   4

Step 3 - Add the multiplicand with the difference of the multiplier **OR** add the multiplier with the difference of the multiplicand.

We will be adding diagonally.

Let us choose to add 22 with 4 to get 26.

Stem 20

$$22 \searrow \times \quad 24 = 26$$
$$2 \quad 4$$

Had you chosen to add 2 with 24 you would have got the same 26.

Stem 20

$$22 \quad \times \searrow 24 = 26$$
$$2 4$$

Since both diagonals give the same result, you can pick the easier of the two.

Step 4 - Multiply the stem number with the result of the addition.

So if we multiply the stem number 20 with the result of the addition 26 we get 520.

Stem 20

$$22 \quad \times \quad 24 = 520$$
$$2 4$$

Step 5 - Multiply the difference of the stem number from the multiplicand with the difference of the stem number from the multiplier.

We multiply 2 x 4 to get 8.

Stem 20

$$22 \quad \times \quad 24 = 520$$
$$2 \quad \times \quad 4 = +8$$

Step 6 - Add the products of the last two steps to get your final answer.

Adding 520 with 8 we get 528.

$$\begin{array}{r} \textbf{Stem 20} \\ 22 \times 24 = 520 \\ 2 \times 4 = + \ 8 \\ \hline 528 \end{array}$$

528 is our final answer.

$$22 \times 24 = 528$$

Let us try another example.

Example 3

Multiply 108 x 96

$$108 \times 96$$

Take a second to apply the technique by yourself as fast as you can. Once you have the answer, you can check the steps below to see if you got your answer right.

Step 1 – Decide a stem number.

We will use 100 as the stem number.

$$\begin{array}{c} \textbf{Stem 100} \\ 108 \times 96 \end{array}$$

Step 2 - Find the difference of the stem number from the multiplicand and the multiplier.

For the multiplicand 108, the difference of 108 – 100 gives us 8.

$$\begin{array}{c} \text{Stem 100} \\ 108 \times 96 \\ 8 \end{array}$$

For the multiplier 96, the difference of 96 – 100 gives us -4.

$$\begin{array}{c} \text{Stem 100} \\ 108 \times 96 \\ 8 \quad\quad -4 \end{array}$$

Step 3 - Add the multiplicand with the difference of the multiplier **OR** add the multiplier with the difference of the multiplicand.

We will be adding diagonally.

Let us choose to add 108 with -4 to get 104.

$$\begin{array}{c} \text{Stem 100} \\ 108 \times 96 = 104 \\ 8 \quad\quad -4 \end{array}$$

Had you chosen to add 96 with 8 you would have got the same 104.

$$\begin{array}{c} \text{Stem 100} \\ 108 \times 96 = 104 \\ 8 \quad\quad -4 \end{array}$$

Since both diagonals give the same result, we will pick the easier of the two.

Step 4 - Multiply the stem number with the result of the addition.

So if we multiply the stem number 100 with the result of the addition 104 we get 100 x 104 = 10400.

        Stem 100
   108 x 96 = 10400
    8   -4

Step 5 - Multiply the difference of the stem number from the multiplicand with the difference of the stem number from the multiplier.

So we multiply 8 x -4 to get -32.

        Stem 100
   108 x 96 = 10400
    8   x -4 = - 32

Step 6 - Add the products of the last two steps to get your final answer.

So adding 10400 with -32 we get the final answer 10368.

        Stem 100
   108 x 96 = 10400
    8   x -4 = - 32
                       10368

When subtracting from 10400 remember the rule the last digits add up to 10 and the other digits add up to 9. Here the last digits 2 in 32 adds up with 8 in 10368 to give 10. The other digit 3 in 32 adds up with 6 in 10368 to give 9.

108 x 96 = 10368

If you got your answer wrong, don't worry. Just revisit the techniques and examples we covered in this chapter.

Read this chapter again if necessary. Then go to the practice section and complete the exercises.

You might have understood the technique, but only practice will make using these techniques effortless and easy.

If you have any questions, you can ask it in the math Q & A section of the community by going **ofpad.com/mathqa** and we will make sure to respond. If you have not become a member of the community yet, watch the video **ofpad.com/communityguide** to get started.

Once you finish practicing, move on to the next section.

Exercises

Download the rich PDFs for these exercises from **ofpad.com/mathexercises**.

01) 97 x 103
02) 101 x 104
03) 96 x 104
04) 49 x 49
05) 53 x 47
06) 48 x 52
07) 17 x 19
08) 18 x 21
09) 18 x 16
10) 37 x 39
11) 36 x 36
12) 36 x 43
13) 33 x 34
14) 30 x 31
15) 33 x 33
16) 22 x 26
17) 24 x 25
18) 29 x 25

Answers

01) 9991
02) 10504
03) 9984
04) 2401
05) 2491
06) 2496

07) 323
08) 378
09) 288
10) 1443

11) 1296
12) 1548
13) 1122
14) 930

15) 1089
16) 572
17) 600
18) 725

Choosing The Stem

Which numbers should you use as stem numbers?

You can use any number as stem number but since our purpose is to make the multiplication problem easy, it is recommended that you try to use 10 or 100 first. If that does not simplify the problem, next try to use 20 or 50.

If you cannot use these numbers, you can use 30 and 40 followed by the other multiple of 10.

Distance From The Stem

So far, we have seen numbers which are closer together. What happens when the numbers are further away from each other?

You can still apply the same technique. The intermediate steps might be a little harder to do but you should easily be able to do the intermediate calculation with the LR Method.

Example 1

For example, let us say you want to multiply 78 x 42.

$$78 \times 42$$

Step 1 – Decide a stem number.

We will use 50 as the stem number since it makes the multiplication easy.

Stem 50
78 x 42

Step 2 - Find the difference of the stem number from the multiplicand and the multiplier.

For the multiplicand 78 the difference of 78 – 50 gives us +28

Stem 50
78 x 42
28

For the multiplier 42 the difference of 42 – 50 gives us -8

Stem 50
78 x 42
28   -8

Step 3 - Add the multiplicand with the difference of the multiplier **OR** add the multiplier with the difference of the multiplicand.

We will be adding diagonally.

You could choose to do the easier addition of 78 with -8 to get 70.

Stem 50
78 x 42 = 70
28   -8

Or you could choose the harder addition of 28 with 42 and you would have got the same 70.

Stem 50

78   x   42   =   70
28   -8

Step 4 - Multiply the stem number with the result of the addition.

So, if we multiply the stem number 50 with the result of the addition 70 we get 50 x 70 = 3500.

Stem 50

78   x   42   =   3500
28   -8

Step 5 - Multiply the difference of the stem number from the multiplicand with the difference of the stem number from the multiplier.

So we multiply +28 x -8 to get -224 using the LR Method.

Stem 50

78   x   42   =   3500
28   x   -8   =   - 224

Step 6 - Add the products of the last two steps to get your final answer.

So adding 3500 with -224 we get 3276. Remember the last digits 4 in 224 and 6 in 3276 add up to 10 and the second 2 in 224 and the 7 in 3276 add up to 9. Also 5 in 3500 adds up with the first 2 and 224 and the 2 in 3276 to give 9.

```
           Stem 50
    78  x   42  =  3500
    28  x   -8  = - 224
                   ────
                   3276
```

3276 is our final answer.

$$78 \times 42 = 3276$$

Let us try another example.

Example 2

Multiply 88 x 68

$$88 \times 68$$

Step 1 – Decide a stem number.

We will use 100 as the stem number since it our preferred stem number.

```
      Stem 100
      88  x  68
```

Step 2 - Find the difference of the stem number from the multiplicand and the multiplier.

For the multiplicand 88 the difference of 88 − 100 gives us -12.

```
      Stem 100
      88  x  68
      -12
```

For the multiplier 68 the difference of 68 − 100 gives us -32.

Stem 100

 88   x   68
-12   -32

If you have trouble with this subtraction, remember that the last digits 8 in 68 and 2 in 32 add up to 10 and the other digits 6 in 68 and 3 in 32 add up to 9.

Step 3 - Add the multiplicand with the difference of the multiplier **OR** add the multiplier with the difference of the multiplicand.

We will be adding diagonally.

You could add of 88 with -32 to get 56.

Stem 100

 88   x   68   = 56
-12   -32

Or you could add of -12 with 68 and you would have got the same 56. Choose the diagonal you find easier.

Stem 100

 88   x   68   = 56
-12   -32

Step 4 - Multiply the stem number with the result of the addition.

So if we multiply the stem number 100 with the result of the addition 56 we get 100 x 56 = 5600.

```
         Stem 100
  88  x   68  = 5600
  -12      -32
```

Step 5 - Multiply the difference of the stem number from the multiplicand with the difference of the stem number from the multiplier.

So we multiply -12 x -32 to get +384. This multiplication might be easier if you use the LR method.

```
         Stem 100
  88  x   68  = 5600
  -12  x  -32 = + 384
```

Step 6 - Add the products of the last two steps to get your final answer.

So, adding 5600 with 384 we get 5984.

```
         Stem 100
  88  x   68  = 5600
  -12  x  -32 = + 384
                 5984
```

5984 is our final answer.

$$88 \times 68 = 5984$$

Let us try another example.

Example 3

Multiply 88 x 92

88 x 92

Take a second to apply the technique by yourself as fast as you can. Once you have the answer, you can check the steps below to see if you got your answer right.

Step 1 – Decide a stem number.

We will use 100 as the stem number since it makes the multiplication easy.

Stem 100
88 x 92

Step 2 - Find the difference of the stem number from the multiplicand and the multiplier.

For the multiplicand 88 the difference of 88 – 100 gives us -12

Stem 100
88 x 92
-12

For the multiplier 92 the difference of 92 – 100 gives us -8

Stem 100
88 x 92
-12   -8

Step 3 - Add the multiplicand with the difference of the multiplier **OR** add the multiplier with the difference of the multiplicand. We will be adding diagonally.

You could choose to do the addition of 88 with -8 to get 80.

$$\begin{array}{c} \text{Stem 100} \\ 88 \quad \times \quad 92 \quad = 80 \\ -12 \qquad\quad -8 \end{array}$$

Or you could choose the addition of -12 with 92 and you would have got the same 80.

$$\begin{array}{c} \text{Stem 100} \\ 88 \quad \times \quad 92 \quad = 80 \\ -12 \qquad\quad -8 \end{array}$$

Step 4 - Multiply the stem number with the result of the addition.

So if we multiply the stem number 100 with the result of the addition 80 we get 8000.

$$\begin{array}{c} \text{Stem 100} \\ 88 \quad \times \quad 92 \quad = 8000 \\ -12 \qquad\quad -8 \end{array}$$

Step 5 - Multiply the difference of the stem number from the multiplicand with the difference of the stem number from the multiplier.

So we multiply -12 x -8 to get +96.

$$\begin{array}{c} \text{Stem 100} \\ 88 \quad \times \quad 92 \quad = 8000 \\ -12 \quad \times \quad -8 \quad = +\ 96 \end{array}$$

Step 6 - Add the products of the last two steps to get your final answer.

So adding 8000 with 96 we get the final answer 8096

Stem 100
$$88 \times 92 = 8000$$
$$-12 \times -8 = +\ \underline{96}$$
$$8096$$

8096 is the answer of the multiplication problem 88 x 92.

88 x 92 = 8096

If you got your answer wrong, don't worry. Just revisit the techniques and examples we covered in this chapter.

Read this chapter again if necessary. Then go to the practice section and complete the exercises.

You might have understood the technique, but only practice will make using these techniques effortless and easy.

If you have any questions, you can ask it in the math Q & A section of the community by going **ofpad.com/mathqa** and we will make sure to respond. If you have not become a member of the community yet, watch the video **ofpad.com/communityguide** to get started.

If you have enjoyed the book so far, do leave a review on Amazon by visiting **ofpad.com/mathbook**.

If you did not enjoy the book so far, and if you have any general suggestions to improve the book or specific feedback for this chapter, do let us know at **ofpad.com/feedback**. If your feedback helps us improve the book even in small ways, we will thank you by sending a free review copy of our next product when it becomes available.

Once you finish practicing, move on to the next section.

Exercises

Download the rich PDFs for these exercises from **ofpad.com/mathexercises**.

01) 98 x 103
02) 82 x 110
03) 91 x 116
04) 49 x 52
05) 35 x 67
06) 41 x 67

07) 5 x 35
08) 19 x 36
09) 9 x 30
10) 36 x 52
11) 26 x 52
12) 22 x 50

13) 27 x 43
14) 21 x 39
15) 18 x 47
16) 16 x 42
17) 21 x 26
18) 6 x 31

Answers

01) 10094
02) 9020
03) 10556
04) 2548
05) 2345
06) 2747

07) 175
08) 684
09) 270
10) 1872
11) 1352
12) 1100

13) 1161
14) 819
15) 846
16) 672
17) 546
18) 186

When to Use Stem Method?

Most find the LR method easier for one-digit multiplication.

When you have to do a 2 by 2 multiplication, the stem method becomes easier.

Applying the stem method becomes easier if the multiplicand and the multiplier are close to each other and if the stem numbers are 10, 100, 20 and 50.

When you have to use another stem, then you might have to use the LR method for the intermediate steps of the stem method.

Chapter 12 - Math Anxiety

When some of us try to do the math, we have butterflies in our stomach. This sometimes also gives us sweaty palms and makes it hard to concentrate.

This phenomenon is called math anxiety, and most of us suffer from it in varying degrees.

You might think you are anxious about math because you are bad at it. But it's often the other way around. You are doing poorly in math because you are anxious about it.

Math anxiety decreases your working memory. Worrying about being able to solve a math problem, eats up working memory leaving less of it available to tackle math itself.

So we can struggle with even the basic math skills that we have otherwise mastered. Relaxing when doing math is therefore essential.

The secrets of doing mental math in this book is going to change the way you do math forever. This is going to make doing math in your head as easy as reading a comic.

But your new math superpower is not going to make you a superhero unless you relax. If you ever feel math anxiety, take a few deep breaths, and you will be able to do the math faster.

Use the knowledge about how your working memory works to create a mindset where you enjoy doing the math. See every math problem as a challenge that will increase your intelligence.

See yourself not only doing math faster but also enjoying yourself while doing it.

Do this, and you will increase your math speed even further.

Chapter 13 - Squaring

In this chapter, we will look at how to square numbers fast. Squaring a number means multiplying a number by itself (example 5 x 5 or 43 x 43). It is usually written as 5^2 or 5^2.

Squaring Numbers Ending With 5

Let us first look at squaring numbers which end with 5 first. We will look at squaring any number next, and it uses the same principle as squaring number ending with 5.

Method

To square a number ending with 5:

Step 1 - Multiply the first digit with the next higher digit to get the first few digits of the answer.

Step 2 - Attach 25 to the first few digits of your answer to get the full answer.

Example 1

Let us look at an example.

Square 65.

```
    6 5
  x 6 5
  ------
```

Step 1 - Multiply the first digit with the next higher digit to get the first few digits of the answer.

So the first digit 6 is multiplied by the next higher digit 7 to give 42.

```
      6 5
    x 6 5
    ─────
    4 2 _ _
```

Step 2 - Attach 25 to the first few digits of your answer to get the full answer.

Attaching 25 to 42 we get 4225.

```
      6 5
    x 6 5
    ─────
    4 2 2 5
```

Your final answer is 4225.

Example 2

Square 45.

```
      4 5
    x 4 5
    ─────
    ─ ─ ─ ─
```

Step 1 - Multiply the first digit with the next higher digit to get the first few digits of the answer.

So the first digit 4 multiplied by the next higher digit 5 gives us 20.

```
      4 5
    x 4 5
    ─────
    2 0 _ _
```

Step 2 - Attach 25 to the first few digits of your answer to get the full answer.

Attaching 25 to 20 we get 2025.

$$\begin{array}{r} 45 \\ \times 45 \\ \hline 2025 \end{array}$$

2025 is your final answer.

Example 3

Square the number 85.

$$\begin{array}{r} 85 \\ \times 85 \\ \hline \end{array}$$

Take a second to apply the technique by yourself as fast as you can. Once you have the answer, you can check the steps below to see if you got your answer right.

Step 1 - Multiply the first digit with the next higher digit to get the first few digits of the answer.

So the first digit 8 is multiplied by the next higher digit 9 to give 72.

$$\begin{array}{r} 85 \\ \times 85 \\ \hline 72\_\_ \end{array}$$

Step 2 - Attach 25 to this number to get the full answer.

```
  85
x 85
----
7225
```

So we have the final answer as 7225.

If you got your answer wrong, don't worry. Just revisit the techniques and examples we covered in this chapter.

If you have any questions, you can ask it in the math Q & A section of the community by going **ofpad.com/mathqa** and we will make sure to respond. If you have not become a member of the community yet, watch the video **ofpad.com/communityguide** to get started.

Squaring Any Number

Numbers that you want to square does not always end with 5. So we will look at the general method of squaring numbers.

Step 1 - Round the number up or down to the nearest multiple of 10.

Step 2 - If you rounded-up, subtract the number by the amount by which you rounded-up. If you rounded-down, add instead of subtract.

Step 3 - Multiply the numbers from the previous two steps.

Step 4 - Square the amount you rounded-up/down and add it to the number from the previous step.

The squaring of numbers ending with 5 is just a simplification of this method.

Example 1

Let us look at an example to understand the method further.

Square the number 23.

$$\begin{array}{r} 23 \\ \times\,23 \\ \hline \end{array}$$

Step 1 - Round the number up or down to the nearest multiple of 10.

So round down 23 to 20.

$$\begin{array}{r} 23 \\ \times\,\underline{23} - 3 = \underline{20} \end{array}$$

Step 2 - If you rounded-up, subtract the number by the amount by which you rounded-up. If you rounded-down, add instead of subtract.

Since we rounded-down, we add the amount we rounded-down, which is 3 to get 26.

$$\begin{array}{r} 23 + 3 = 26 \\ \times\,\underline{23} - 3 = \underline{20} \end{array}$$

Step 3 - Multiply the numbers from the previous two steps. Remember to multiply from left to right.

Multiply 26 with 20 to get 520.

$$23 + 3 = 26$$
$$\underline{\times\ 23} \quad \underline{-\ \ 3} = \underline{\times\ 20}$$
$$520$$

Step 4 - Square the amount you rounded-up/down and add it to the number from the previous step.

We rounded-down 23 by 3. Squaring 3, we get 9.

$$23 + 3 = 26$$
$$\underline{\times\ 23} \quad \underline{-\times 3} = \underline{\times\ 20}$$
$$9 \qquad 520$$

Adding 9 to 520 we get 529.

$$23 + 3 = 26$$
$$\underline{\times\ 23} \quad \underline{-\times 3} = \underline{\times\ 20}$$
$$9\ +\ 520 = 529$$

And you have the final answer of the square of 23 as 529.

$$23$$
$$\underline{\times\ 23}$$
$$529$$

Example 2

Square the number 48.

$$48$$
$$\underline{\times\ 48}$$
$$----$$

Step 1 - Round the number up or down to the nearest multiple of 10.

So round up 48 to 50.

$$\begin{array}{r} 4\,8 \\ \times \underline{4\,8 + 2 = 50} \end{array}$$

Step 2 - If you rounded-up, subtract the number by the amount by which you rounded-up. If you rounded-down, add instead of subtract.

Since we rounded-up, we subtract the amount we rounded-up, which is 2 to get 46.

$$\begin{array}{r} 4\,8 - 2 = 46 \\ \times \underline{4\,8 + 2 = 50} \end{array}$$

Step 3 - Multiply the numbers from the previous two steps.

Multiply 46 with 50 to get 2300. Remember to multiply from left to right.

$$\begin{array}{r} 4\,8 - 2 = 46 \\ \times \underline{4\,8 + 2 =\times 50} \\ 2300 \end{array}$$

Step 4 - Square the amount you rounded-up/down and add it to the number from the previous step.

We rounded-up 48 by 2. Squaring 2, we get 4.

$$\begin{array}{r} 4\,8 - 2 = 46 \\ \times \underline{4\,8 + \times 2 =\times 50} \\ 4 2300 \end{array}$$

Adding it to 2300, we get 2304.

$$\begin{array}{r} 48 \\ \times\ 48 \\ \hline 4 \end{array} \quad \begin{array}{r} -\ 2 = 46 \\ +\ \times 2 = \times 50 \\ \hline +2300 = 2304 \end{array}$$

So the square of 48 is 2304 which is your final answer.

$$\begin{array}{r} 48 \\ \times 48 \\ \hline 2304 \end{array}$$

Example 3

Square the number 64.

$$\begin{array}{r} 64 \\ \times 64 \\ \hline ---- \end{array}$$

Take a second to apply the technique by yourself as fast as you can. Once you have the answer, you can check the steps below to see if you got your answer right.

Step 1 - Round the number up or down to the nearest multiple of 10.

So round down 64 to 60.

$$\begin{array}{r} 64 \\ \times\ 64\ -\ 4 = 60 \end{array}$$

Step 2 - If you rounded-up, subtract the number by the amount by which you rounded-up. If you rounded-down, add instead of subtract.

Since we rounded-down, we add the amount we rounded-down, which is 4 to get 68.

$$\begin{array}{r} 64 + 4 = 68 \\ \times\ 64 - 4 = 60 \\ \hline \end{array}$$

Step 3 - Multiply the numbers from the previous two steps. Remember to multiply from left to right.

Multiply 68 with 60 to get 4080.

$$\begin{array}{r} 64 + 4 = 68 \\ \times\ 64 - 4 =\times 60 \\ \hline 4080 \end{array}$$

Step 4 - Square the amount you rounded-up/down and add it to the number from the previous step.

We rounded-down 64 by 4. Squaring 4, we get 16.

$$\begin{array}{r} 64 + 4 = 68 \\ \times\ 64 - \times 4 =\times 60 \\ \hline 16 \quad 4080 \end{array}$$

Adding it to 4080, we get 4096.

$$\begin{array}{r} 64 + 4 = 68 \\ \times\ 64 - \times 4 =\times 60 \\ \hline 16 + 4080 = 4096 \end{array}$$

You have now got the square of 64 as 4096.

$$\begin{array}{r} 64 \\ \times 64 \\ \hline 4096 \end{array}$$

If you got your answer wrong, don't worry. Just revisit the techniques and examples we covered in this chapter.

Read this chapter again if necessary. Then go to the practice section and complete the exercises.

You might have understood the technique, but it will take practice before the technique becomes second nature to you. Only practice will make using these techniques effortless and easy.

If you have any questions, you can ask it in the math Q & A section of the community by going **ofpad.com/mathqa** and we will make sure to respond. If you have not become a member of the community yet, watch the video **ofpad.com/communityguide** to get started.

If you have enjoyed the book so far, do leave a review on Amazon by visiting **ofpad.com/mathbook**.

If you did not enjoy the book so far, and if you have any general suggestions to improve the book or specific feedback for this chapter, do let us know at **ofpad.com/feedback**. If your feedback helps us improve the book even in small ways, we will thank you by sending a free review copy of our next product when it becomes available.

Once you finish practicing, move on to the next section.

Exercises

Download the rich PDFs for these exercises from **ofpad.com/mathexercises**.

01) 35
   x 35

02) 73
   x 73

03) 46
   x 46

04) 97
   x 97

05) 24
   x 24

06) 83
   x 83

07) 56
   x 56

08) 71
   x 71

09) 39
   x 39

10) 63
   x 63

11) 83
   x 83

12) 28
   x 28

13) 85
   x 85

14) 15
   x 15

15) 53
   x 53

16) 330
   x 330

17) 745
   x 745

18) 515
   x 515

Answers

01) 1,225
02) 5,329
03) 2,116
04) 9,409
05) 576
06) 6,889
07) 3,136
08) 5,041
09) 1,521
10) 3,969
11) 6,889
12) 784
13) 7,225
14) 225

15) 2,809
16) 108,900
17) 555,025
18) 265,225

Chapter 14 - The Bridge Method

In this chapter, we will look at how to do fast multiplication using the bridge method. You will be able to multiply two-digit and three-digit numbers together using this method.

To multiply using the bridge method:

Step 1 - Multiply the first number of the multiplicand with the first number of the multiplier and put the result as the left-hand number of the answer.

Step 2 - Multiply the outside pairs and multiply the inside pairs.

Step 3 - Add the products of the outside and inside pairs together to get the next figure of the answer.

Step 4 - Multiply the last number of the multiplicand with the second number of the multiplier and put the result as the right-hand number of the answer.

Tip - Use your forefinger and middle finger to pace through the multiplication problem. Place them on the screen/paper, so you keep track of which numbers you are calculating.

Let us look at an example.

Example 1

Multiply 32 by 13.

$$32 \times 13$$

Step 1 - Multiply the first number of the multiplicand with the first number of the multiplier and put the result as the left-hand number of the answer.

Multiply 3 with 1 to give 3.

$$3\ 2\ \text{x}\ 1\ 3\ =\ 3\ \_\ \_$$

Step 2 - Multiply the outside pairs and multiply the inside pairs.

So multiplying the outside pairs 3 and 3, we get 9.

Multiplying the inside pairs 2 and 1, we get 2.

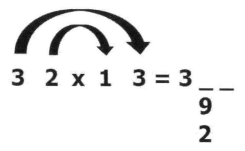

Step 3 - Add the products of the outside and inside pairs together to get the next figure of the answer.

Adding 9 and 2, we get 11.

$$3\ 2\ \text{x}\ 1\ 3 = 3\ \_\ \_$$

$$\begin{array}{r} 9 \\ +2 \\ \hline 11 \end{array}$$

Carry over the one so that 3 becomes 4.

$$3\ 2\ \text{x}\ 1\ 3 = 4\ 1\ \_$$

Step 4 - Multiply the second number of the multiplicand with the second number of the multiplier and put the result as the right-hand number of the answer.

So multiplying 2 and 3 we get 6 which is the last number of the answer.

$$3\ 2\ \text{x}\ 1\ 3 = 4\ 1\ 6$$

The final answer is 416.

Example 2

Multiply 323 with 13.

Use your forefinger and middle finger to pace through the multiplication problem. Place them on the screen/paper, so you keep track of which numbers you are calculating.

$$3\ 2\ 3\ \text{x}\ 1\ 3$$

Step 1 - Multiply the first number of the multiplicand with the first number of the multiplier and put the result as the left-hand number of the answer.

Multiply 3 with 1 to give 3.

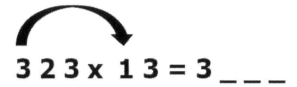

Step 2 - Multiply the outside pairs and multiply the inside pairs.

So multiplying the outside pairs 3 and 3, we get 9.

Multiplying the inside pairs 2 and 1, we get 2.

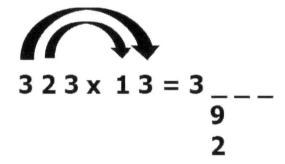

Step 3 - Adding the products of the outside and inside pairs together we get the next figure of the answer.

Adding 2 and 9 we get 11.

$$3\,2\,3 \times 1\,3 = 3\,\_\,\_\,\_$$
$$9$$
$$+2$$
$$\overline{11}$$

Carry over the one so that 3 becomes 4.

$$3\,2\,3 \times 1\,3 = 4\,1\,\_\,\_$$

Repeat the steps for the next set of digits.

Step 2 - Multiply the outside pairs and multiply the inside pairs.

So multiplying the outside pairs 2 and 3, we get 6.

Multiplying the inside pairs 3 and 1, we get 3.

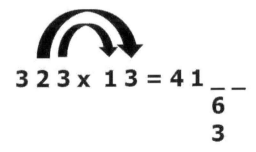

$$3\,2\,3 \times 1\,3 = 4\,1\,\_\,\_$$
$$6$$
$$3$$

Step 3 - Add the products of the outside and inside pairs together to get the next figure of the answer.

Adding 6 and 3, we get 9.

$$323 \times 13 = 41\_\_$$
$$6$$
$$+3$$
$$\overline{9}$$

Step 4 - Multiply the last number of the multiplicand with the second number of the multiplier and put the result as the right-hand number of the answer.

So multiplying 3 and 3 we get 9 which is the last number of the answer.

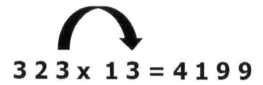

$$323 \times 13 = 4199$$

The answer is 4199.

Example 3

Multiply 323 by 132.

$$323 \times 132$$

Take a second to apply the technique by yourself as fast as you can. Once you have the answer, you can check the steps below to see if you got your answer right. Remember to use your forefinger and middle finger to pace through the multiplication problem.

Step 1 - Multiply the first number of the multiplicand with the first number of the multiplier and put the result as the left-hand number of the answer.

Multiply 3 with 1 to give 3.

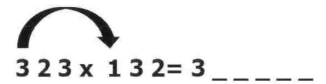

3 2 3 x 1 3 2 = 3 _ _ _ _ _

Step 2 - Multiply the outside pairs and multiply the inside pairs.

So multiplying the outside pairs 3 and 3 we get 9.

Multiplying the inside pairs 2 and 1 we get 2.

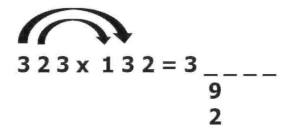

$$3\ 2\ 3 \times 1\ 3\ 2 = 3\ \_\ \_\ \_\ \_$$
$$9$$
$$2$$

Step 3 - Add the products of the outside and inside pairs together to get the next figure of the answer.

Adding 9 and 2 we get 11.

$$3\ 2\ 3 \times 1\ 3\ 2 = 3\ \_\ \_\ \_\ \_$$
$$9$$
$$\underline{+2}$$
$$11$$

Carry over the 1 so the 3 becomes 4.

$$3\ 2\ 3 \times 1\ 3\ 2 = 4\ 1\ \_\ \_\ \_$$

Repeat the steps for the next set of digits

Step 2 - Multiply the outside pairs and multiply the inside pairs.

   a) Multiplying the outside pairs 3 and 2 we get 6.
   b) Multiplying the inside pairs 2 and 3 we get 6.
   c) Multiplying the last set of inner pairs 3 and 1 we get 3.

$$3\ 2\ 3 \times 1\ 3\ 2 = 4\ 1\ \_\ \_\ \_$$
$$6$$
$$6$$
$$\underline{+3}$$

Step 3 - Add the products of the outside and inside pairs together to get the next figure of the answer.

Adding 6 with 6 and then adding it with 3 we get 15.

$$3\ 2\ 3 \times 1\ 3\ 2 = 4\ 1\ \_\ \_\ \_$$
$$6$$
$$6$$
$$\underline{+3}$$
$$15$$

Carry over the one, so the 1 becomes 2.

$$3\ 2\ 3 \times 1\ 3\ 2 = 4\ 2\ 5\ \_\ \_$$

Repeat the steps for the next set of digits.

Step 2 - Multiply the outside pairs and multiply the inside pairs.

So multiplying the outside pairs 2 and 2 we get 4.

Multiplying the inside pairs 3 and 3 we get 9.

$$3\ 2\ 3 \times 1\ 3\ 2 = 4\ 2\ 5\ \_\ \_$$
$$4$$
$$\underline{+9}$$

Step 3 - Add the products of the outside and inside pairs together to get the next figure of the answer.

Adding 4 with 9 we get 13.

$$3\ 2\ 3 \times 1\ 3\ 2 = 4\ 2\ 5\ \_\ \_$$
$$4$$
$$\underline{+9}$$
$$13$$

Carry over the one, so the 5 becomes 6.

$$3\ 2\ 3 \times 1\ 3\ 2 = 4\ 2\ 6\ 3\ \_$$

Step 4 - Multiply the last number of the multiplicand with the last number of the multiplier and put the result as the right-hand number of the answer.

So multiplying 3 and 2 we get 6 which is the last number of the answer.

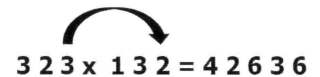

$$3\ 2\ 3 \times 1\ 3\ 2 = 4\ 2\ 6\ 3\ 6$$

The answer is 42,636.

If you got your answer wrong, don't worry. Just revisit the techniques and examples we covered in this chapter.

Read this chapter again if necessary. Then go to the practice section and complete the exercises.

You might have understood the technique, but it will take practice before the technique becomes second nature to you.

Only practice will make using these techniques effortless and easy.

If you have any questions, you can ask it in the math Q & A section of the community by going **ofpad.com/mathqa** and we will make sure to respond. If you have not become a member of the community yet, watch the video **ofpad.com/communityguide** to get started.

If you have enjoyed the book so far, do leave a review on Amazon by visiting **ofpad.com/mathbook**.

If you did not enjoy the book so far, and if you have any general suggestions to improve the book or specific feedback for this chapter, do let us know at **ofpad.com/feedback**. If your feedback helps us improve the book even in small ways, we will thank you by sending a free review copy of our next product when it becomes available.

Once you finish practicing, move on to the next section.

Exercises

Download the rich PDFs for these exercises from **ofpad.com/mathexercises**.

01) 38 x 74
02) 63 x 63
03) 14 x 22
04) 12 x 57
05) 26 x 95
06) 25 x 87
07) 917 x 13
08) 140 x 66
09) 389 x 60
10) 852 x 94
11) 459 x 92
12) 687 x 89
13) 368 x 349
14) 106 x 783
15) 312 x 642
16) 430 x 152
17) 970 x 639
18) 553 x 963

Answers

01) 2,812
02) 3,969
03) 308
04) 684
05) 2,470
06) 2,175
07) 11,921
08) 9,240
09) 23,340
10) 80,088
11) 42,228
12) 61,143
13) 128,432
14) 82,998
15) 200,304
16) 65,360
17) 619,830
18) 532,539

Chapter 15 - Vitruvian Man Method

It is called the Vitruvian Man method because the visualisation methods resemble that of a Vitruvian man. The steps in the method are same as the bridge method. The only thing which is different is the mental visualisation.

Use your forefinger and middle finger to pace through the multiplication problem.

Example 1

Multiply 32 by 13.

$$\begin{array}{r} 3\ 2 \\ \times 1\ 3 \\ \hline -\ -\ - \end{array}$$

Let us apply the new method of visualisation.

Use your forefinger and middle finger to pace through the multiplication problem.

Step 1 - Multiply the first number of the multiplicand with the first number of the multiplier and put the result as the left-hand number of the answer.

Multiply 3 with 1 to give 3.

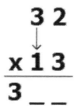

Step 2 - Multiply the outside pairs and multiply the inside pairs.

So multiplying the outside pairs 3 and 3 we get 9.

Multiplying the inside pairs 2 and 1 we get 2.

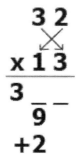

Step 3 - Add the products of the outside and inside pairs together to get the next figure of the answer

Adding 2 and 9, we get 11.

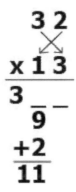

Carry over the one so that 3 becomes 4.

```
   3 2
 x 1 3
 ─────
 4 1 _
```

Step 4 - Multiply the second number of the multiplicand with the second number of the multiplier and put the result as the right-hand number of the answer.

So multiplying 2 and 3 we get 6 which is the last number of the answer.

```
   3 2
     ↓
 x 1 3
 ─────
 4 1 6
```

The answer is 416.

Example 2

Multiply 323 with 13.

$$\begin{array}{r}323\\ \times 13\\ \hline \_\_\_\_\end{array}$$

Apply the new method of visualisation.

Use your forefinger and middle finger to pace through the multiplication problem.

Step 1 - Multiply the first number of the multiplicand with the first number of the multiplier and put the result as the left-hand number of the answer.

Multiply 3 with 1 to give 3.

$$\begin{array}{r}323\\ \times 13\\ \hline 3\_\_\_\end{array}$$

Step 2 - Multiply the outside pairs and multiply the inside pairs.

So multiplying the outside pairs 3 and 3 we get 9.

Multiplying the inside pairs 2 and 1 we get 2.

```
  3 2 3
×   1 3
─────────
3 _ _ _
  9
 +2
```

Step 3 - Add the products of the outside and inside pairs together to get the next figure of the answer.

Adding 2 and 9, we get 11.

```
  3 2 3
×   1 3
─────────
3 _ _ _
  9
 +2
─────
 11
```

Carry over the one so that 3 becomes 4.

```
  3 2 3
×   1 3
─────────
4 1 _ _
```

Repeat the steps for the next set of digits.

Step 2 - Multiply the outside pairs and multiply the inside pairs.

So multiplying the outside pairs 2 and 3, we get 6.

Multiplying the inside pairs 3 and 1, we get 3.

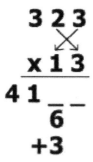

Step 3 - Add the products of the outside and inside pairs together to get the next figure of the answer

Adding 6 and 3, we get 9.

```
      3 2 3
       ⤫
    x 1 3
    ─────
    4 1 _ _
        6
       +3
       ───
        9
```

So 9 becomes the next digit of the answer.

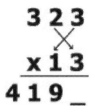

Step 4 - Multiply the last number of the multiplicand with the second number of the multiplier and put the result as the right-hand number of the answer.

So multiplying 3 and 3 we get 9 which is the last number of the answer.

```
  3 2 3
     ↓
  x 1 3
  ─────
  4 1 9 9
```

The answer is 4199.

Example 3

Multiply 323 by 132

Apply the new method of visualisation.

```
  3 2 3
  x 1 3 2
  ───────
  _ _ _ _ _
```

Take a second to apply the technique by yourself as fast as you can. Once you have the answer, you can check the steps below to see if you got your answer right. Remember to use your forefinger and middle finger to pace through the multiplication problem

Step 1 - Multiply the first number of the multiplicand with the first number of the multiplier and put the result as the left-hand number of the answer.

Multiply 3 with 1 to give 3.

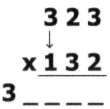

Step 2 - Multiply the outside pairs and multiply the inside pairs.

So multiplying the outside pairs 3 and 3 we get 9.

Multiplying the inside pairs 2 and 1 we get 2.

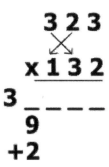

Step 3 - Add the products of the outside and inside pairs together to get the next figure of the answer.

Adding 9 and 2, we get 11.

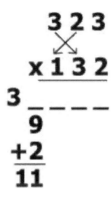

Carry over the 1 so the 3 becomes 4.

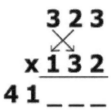

Repeat the steps for the next set of digits.

Step 2 - Multiply the outside pairs and multiply the inside pairs.

    a) So multiplying the outside pairs 3 and 2 we get 6.
    b) Multiplying the inside pairs 2 and 3 we get 6.
    c) Multiply the last set of inner pairs 3 and 1 to give 3.

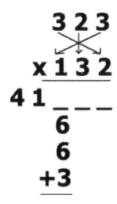

Step 3 - Add the products of the outside and inside pairs together to get the next figure of the answer.

Adding 6 with 6 and then adding it with 3 we get 15.

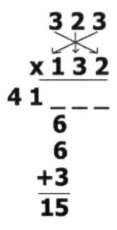

Carry over the one, so the 1 becomes 2.

```
    3 2 3
  x 1 3 2
  ─────────
  4 2 5 _ _
```

Repeat the steps for the next set of digits.

Step 2 - Multiply the outside pairs and multiply the inside pairs.

$$\begin{array}{r} 3\,2\,3 \\ \times\,1\,3\,2 \\ \hline 4\,2\,5\,\_\,\_ \end{array}$$

So multiplying the outside pairs 2 and 2 we get 4.

Multiplying the inside pairs 3 and 3 we get 9.

$$\begin{array}{r} 3\,2\,3 \\ \times\,1\,3\,2 \\ \hline 4\,2\,5\,\_\,\_ \\ 4 \\ +9 \\ \hline \end{array}$$

Step 3 - Add the products of the outside and inside pairs together to get the next figure of the answer.

Adding 4 with 9 we get 13.

$$\begin{array}{r} 3\,2\,3 \\ \times\,1\,3\,2 \\ \hline 4\,2\,5\,\_\,\_ \\ 4 \\ +9 \\ \hline 13 \end{array}$$

Carry over the one, so the 5 becomes 6.

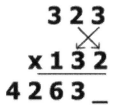

```
  3 2 3
x 1 3 2
———————
  4 2 6 3 _
```

Step 4 - Multiply the last number of the multiplicand with the last number of the multiplier and put the result as the right-hand number of the answer.

```
  3 2 3
      ↓
x 1 3 2
———————
  4 2 6 3 _
```

So multiplying 3 and 2 we get 6 which is the last number of the answer.

```
  3 2 3
      ↓
x 1 3 2
———————
  4 2 6 3 6
```

The answer is 42,636.

If you got your answer wrong, don't worry. Just revisit the techniques and examples we covered in this chapter.

Read this chapter again if necessary. Then go to the practice section and complete the exercises.

You might have understood the technique, but it will take practice before the technique becomes second nature to you.

If you have any questions, you can ask it in the math Q & A section of the community by going **ofpad.com/mathqa** and we will make sure to respond. If you have not become a member of the community yet, watch the video **ofpad.com/communityguide** to get started.

If you have enjoyed the book so far, do leave a review on Amazon by visiting **ofpad.com/mathbook**.

If you did not enjoy the book so far, and if you have any general suggestions to improve the book or specific feedback for this chapter, do let us know at **ofpad.com/feedback**. If your feedback helps us improve the book even in small ways, we will thank you by sending a free review copy of our next product when it becomes available.

Once you finish practicing, move on to the next section.

Exercises

Download the rich PDFs for these exercises from **ofpad.com/mathexercises**.

```
01)  97        04) 421        07) 366
   x  79         x   63         x   68

02)  22        05) 692        08) 139
   x  21         x   38         x   25

03)  41        06) 897        09) 458
   x  32         x   40         x   78
```

10) 306 x 56	13) 252 x 881	16) 865 x 115
11) 874 x 42	14) 582 x 473	17) 396 x 487
12) 422 x 80	15) 406 x 926	18) 699 x 244

Answers

01) 7,663
02) 462
03) 1,312
04) 26,523
05) 26,296
06) 35,880
07) 24,888
08) 3,475
09) 35,724
10) 17,136
11) 36,708
12) 33,760
13) 222,012
14) 275,286
15) 375,956
16) 99,475
17) 192,852
18) 170,556

Chapter 16 - UT Method

UT in UT Method is short for Units and Tens. UT Method is another method of multiplication and you will be able to do large multiplication problems fast.

UT Method Concepts

There are a few concepts and definition you will have to understand before we apply the UT method:

1) A digit is a one figure number (e.g. 4, 2, 0).
2) Multiplying a digit by a digit will give you a one figure or two figure number but never longer (The longest number is 9 x 9 = 81).
3) Sometimes a digit multiplied by a digit will give a one figure number. In such cases for the UT method, we treat it as a two-digit number by adding a 0 in front of it (e.g. 2 x 3 = 06).
4) The left-hand digit is the tens digit (T) and the right-hand digit is the units digit (U) (e.g. In 23, 2 is T and 3 is U).
5) We will use either U or T of a number but never both in UT method.

Let us do a few exercises to reinforce what you have learnt.

Practice Exercise

Find the U of the following multiplication problems. Speed is very important, and you should just see the U without thinking about the T.

Find the U of the following numbers

$$5 \times 3 =$$
$$6 \times 4 =$$
$$9 \times 3 =$$
$$5 \times 2 =$$
$$6 \times 1 =$$
$$7 \times 4 =$$

Here are the answers:

$$5 \times 3 = 5 \text{ (U)}$$
$$6 \times 4 = 4 \text{ (U)}$$
$$9 \times 3 = 7 \text{ (U)}$$
$$5 \times 2 = 0 \text{ (U)}$$
$$6 \times 1 = 6 \text{ (U)}$$
$$7 \times 4 = 8 \text{ (U)}$$

Now let us do the same exercise to find the value of T. Speed is very important and you should just see the T without thinking about the U.

Find the T of the following numbers:

5 x 3 =

6 x 4 =

9 x 3 =

5 x 2 =

6 x 1 =

7 x 4 =

Here are the answers:

5 x 3 = 1 (T)

6 x 4 = 2 (T)

9 x 3 = 2 (T)

5 x 2 = 1 (T)

6 x 1 = 0 (T)

7 x 4 = 2 (T)

Pair Products

Before we look at the UT method, we must understand what a pair product is.

To find the pair product:

Step 1 - We use the multiplier to multiply each digit of the multiplicand separately.

Step 2 - We then take the U of the product of the left-hand digit of the multiplicand and the T of the product of the right-hand digit of the multiplicand.

Step 3 - We then add U and T together to get the pair product.

Let us look at an example to illustrate the method

Example 1

Let us find the pair product of 48 x 6.

$$\begin{array}{r}48\\\underline{\times\ 6}\end{array}$$

Step 1 - We use the multiplier to multiply each digit of the multiplicand separately.

So we multiply 4 by 6 to give 24, and we multiply 8 by 6 to give 48.

$$\begin{array}{r}48\\\underline{\times\ 6}\\24+48\end{array}$$

Step 2 - We then take the U of the product of the left-hand digit of the multiplicand and the T of the product of the right-hand digit of the multiplicand.

U of the first number is 4 and T of the second number is 4.

$$\begin{array}{r}48\\ \times\,6\\ \hline 2\mathbf{4}+\mathbf{4}8\end{array}$$

Step 3 - We then add U and T together to get the pair product.

Adding 4 and 4 we get 8 which is the pair product.

$$\begin{array}{r}\mathbf{48}\\ \mathbf{\times\,6}\\ \hline 2\mathbf{4}+\mathbf{4}8\\ \hline \mathbf{8}\end{array}$$

The numbers which are not in **bold** are there just for your understanding. When you are visualising in your mind, you should not think about the numbers which are not in **bold**. Your focus should just be on the U or the T and adding them to get the pair product.

Example 2

Let us find the pair product of 48 x 5.

$$\begin{array}{r}\mathbf{48}\\ \mathbf{\times\,5}\\ \hline \end{array}$$

Step 1 - We use the multiplier to multiply each digit of the multiplicand separately.

So we multiply 4 by 5 to give 20, and we multiply 8 by 5 to give 40.

$$\begin{array}{r} 48 \\ \underline{\times\, 5} \\ 20 + 40 \end{array}$$

Step 2 - We then take the U of the product of the left-hand digit of the multiplicand and the T of the product of the right-hand digit of the multiplicand.

U of the first number is 0 and T of the second number is 4.

$$\begin{array}{r} 48 \\ \underline{\times\, 5} \\ 2\mathbf{0} + \mathbf{4}0 \end{array}$$

Step 3 - We then add U and T together to get the pair product.

Adding 0 and 4, we get 4 which is the pair product.

$$\begin{array}{r} 48 \\ \underline{\times\, 5} \\ \underline{2\mathbf{0} + \mathbf{4}0} \\ 4 \end{array}$$

Example 3

Let us find the pair product of 41 x 5.

$$\begin{array}{r} 41 \\ \underline{\times\, 5} \end{array}$$

Step 1 - We use the multiplier to multiply each digit of the multiplicand separately.

So we multiply 4 by 5 to give 20, and we multiply 1 by 5 to give 05.

$$\begin{array}{r} 41 \\ \times\, 5 \\ \hline 20 + 05 \end{array}$$

Step 2 - We then take the U of the product of the left-hand digit of the multiplicand and the T of the product of the right-hand digit of the multiplicand.

U of the first number is 0 and T of the second number is 0.

$$\begin{array}{r} 41 \\ \times\, 5 \\ \hline 2\underline{0} + \underline{0}5 \end{array}$$

Step 3 - We then add U and T together to get the pair product.

Adding 0 and 0, we get 0 which is the pair product.

$$\begin{array}{r} 41 \\ \times\, 5 \\ \hline 2\underline{0} + \underline{0}5 \\ 0 \end{array}$$

Example 4

Find the pair product of 28 by 4.

$$\begin{array}{r}28\\ \underline{\times\ 4}\end{array}$$

Take a second to apply the technique by yourself as fast as you can. Once you have the answer, you can check the steps below to see if you got your answer right.

Step 1 - We use the multiplier to multiply each digit of the multiplicand separately.

So we multiply 2 by 4 to give 08, and we multiply 8 by 4 to give 32.

$$\begin{array}{r}28\\ \underline{\times\ 4}\\ 08 + 32\end{array}$$

Step 2 - We then take the U of the product of the left-hand digit of the multiplicand and the T of the product of the right-hand digit of the multiplicand.

U of the first number is 8 and T of the second number is 3.

$$\begin{array}{r}28\\ \underline{\times\ 4}\\ 0\underline{8} + \underline{3}2\end{array}$$

Step 3 - We then add U and T together to get the pair product.

Adding 8 and 3, we get 11 which is the pair product.

```
    28
   x 4
  08 + 32
    11
```

If you got your answer wrong, don't worry. Just revisit the techniques and examples we covered in this chapter.

If you have any questions, you can ask it in the math Q & A section of the community by going **ofpad.com/mathqa** and we will make sure to respond. If you have not become a member of the community yet, watch the video **ofpad.com/communityguide** to get started.

Tips

When applying the UT method:

a) Think of single-digit products as if a 0 is attached in front of it (ex. 3 x 2 = 06).
b) Sometimes when you add U + T, you get a two-digit number. In such cases, you carry over the one, like how you did in previous methods.
c) Visualize the numbers in your mind and make the pair products.
d) When you feel like you are arriving at the answer without focusing on the steps, the process has become truly automatic.
e) Think about only the U or the T and not the number you are dropping. It will save you time as you do the problem.

Applying UT Method

The rule is simple:

Step 1 – First add the same number of zeros in front of the multiplicand as the number of digits in the multiplier.

Step 2 - Each pair product is a single digit of the answer. In case of two-digit multipliers, the sum of the two pair products is a single digit of the answer.

Let us look at an example.

Example 1

Multiply 4312 by 4

$$4\ 3\ 1\ 2 \quad x\ 4$$

Step 1 - Add the same number of zeros as the number of digits in the multiplier.

$$0\ 4\ 3\ 1\ 2 \quad x\ 4$$

Step 2 - Each pair product is a single digit of the answer.

Use your middle and forefinger to keep track of whether to calculate U or the T.

Pair product of 0 and 4 is 0 + 1 = 1 (0<u>0</u> + <u>1</u>6).

```
       U T
    0 4 3 1 2    x 4
    1 _ _ _ _
```

Next, the pair product of 4 and 3 is 6 + 1 = 7 (1<u>6</u> + <u>1</u>2).

```
     U T
  0 4 3 1 2    x 4
  ─────────
  1 7 _ _ _
```

Next, the pair product of 3 and 1 is 2 + 0 = 2 (1<u>2</u> + <u>0</u>4).

```
     U T
  0 4 3 1 2    x 4
  ─────────
  1 7 2 _ _
```

Next, the pair product of 1 and 2 is 4 + 0 = 4 (0<u>4</u> + <u>0</u>8).

```
     U T
  0 4 3 1 2    x 4
  ─────────
  1 7 2 4 _
```

Next, the pair product of 2 and blank is 8 + _ = 8 (0<u>8</u> + <u>-</u>-).

```
     U T
  0 4 3 1 2    x 4
  ─────────
  1 7 2 4 8
```

Now you have the answer 17,248.

The UT Method is not very useful for one-digit multiplier. You could just use the LR method for one-digit multipliers.

However, it really speeds up the calculation for two-digit multipliers.

Example 2

Let us now try a two-digit multiplier. Multiply 4312 by 42.

$$4312 \quad \times 42$$

Step 1 - Add the same number of zeros as the number of digits in the multiplier.

$$004312 \quad \times 42$$

Step 2 - Sum of two pair products is a single digit of the answer.

Use your middle and forefinger to keep track of whether to calculate U or the T. Place each finger in the middle of two pairs.

Using 2 in 42, the pair product of 0 and 0 is 0 + 0 = 0 (U_2T_2 = 0<u>0</u> + <u>0</u>0).
Using 4 in 42, the pair product of 0 and 4 is 0 + 1 = 1 (U_4T_4 = 0<u>0</u> + <u>1</u>6).

Adding the two pair products 0 + 1 together gives us 1.

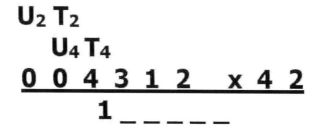

We move to the next two digits.

Using 2 in 42, the pair product of 0 and 4 is 0 + 0 = 0 (U_2T_2 = 0<u>0</u> + <u>0</u>8).
Using 4 in 42, the pair product of 4 and 3 is 6 + 1 = 7 (U_4T_4 = 1<u>6</u> + <u>1</u>2).

Adding the two pair products 0 + 7 together gives us 7.

$$
\begin{array}{r}
U_2\ T_2 \\
U_4\ T_4 \\
0\ 0\ 4\ 3\ 1\ 2 \quad \times\ 4\ 2 \\
\hline
1\ 7\ \_\ \_\ \_\ \_
\end{array}
$$

We move to the next two digits.

Using 2 in 42, the pair product of 4 and 3 is 8 + 0 = 8 (U_2T_2 = 0<u>8</u> + <u>0</u>6).
Using 4 in 42, the pair product of 3 and 1 is 2 + 0 = 2 (U_4T_4 = 1<u>2</u> + <u>0</u>4).

Adding the two pair products 8 + 2 together gives us 10.

$$
\begin{array}{r}
U_2\ T_2 \\
U_4\ T_4 \\
0\ 0\ 4\ 3\ 1\ 2 \quad \times\ 4\ 2 \\
\hline
1\ 7\ \_\ \_\ \_\ \_ \\
1\ 0
\end{array}
$$

Carrying over the 1, the 7 becomes 8.

$$\begin{array}{r} U_2\ T_2 \\ U_4\ T_4 \\ \underline{0\ 0\ 4\ 3\ 1\ 2\quad \times\ 4\ 2} \\ 1\ 8\ 0\ \underline{\ }\ \underline{\ }\ \underline{\ } \end{array}$$

We move to the next two digits.

Using 2 in 42, the pair product of 3 and 1 is 6 + 0 = 6 ($U_2 T_2$ = 0$\underline{6 + 0}$2).

Using 4 in 42, the pair product of 1 and 2 is 4 + 0 = 4 ($U_4 T_4$ = 0$\underline{4 + 0}$8).

Adding the two pair products 6 + 4 together gives us 10.

$$\begin{array}{r} U_2\ T_2 \\ U_4\ T_4 \\ \underline{0\ 0\ 4\ 3\ 1\ 2\quad \times\ 4\ 2} \\ 1\ 8\ 0\ \underline{\ }\ \underline{\ }\ \underline{\ } \\ +\ 1\ 0\quad\quad \end{array}$$

Carry over the 1, so the 0 becomes 1.

$U_2\ T_2$
$U_4\ T_4$

```
0 0 4 3 1 2    x 4 2
      1 8 1 0 _ _
```

We move to the next two digits.

Using 2 in 42, the pair product of 1 and 2 is 2 + 0 = 2 (U_2T_2 = 0<u>2</u> + 0<u>4</u>).

Using 4 in 42, the pair product of 2 and blank is 8 + _ = 8 (U_4T_4 = 0<u>8</u> + --).

Adding the two pair products together 2 + 8 gives us 10

$U_2\ T_2$
$U_4\ T_4$

```
0 0 4 3 1 2    x 4 2
      1 8 1 0 _ _
            +1 0
```

Carrying over the 1, the 0 becomes 1.

$$U_2\ T_2$$
$$U_4\ T_4$$
$$\underline{0\ 0\ 4\ 3\ 1\ 2\quad \times 4\ 2}$$
$$1\ 8\ 1\ 1\ 0\ \_$$

We move to the next two digits.

Using 2 in 42, the pair product of 2 and blank is 4 + _ = 4 ($U_2 T_2$ = 0$\underline{4 + \text{--}}$) and that becomes the last digit of the answer.

$$U_2\ T_2$$
$$U_4\ T_4$$
$$\underline{0\ 0\ 4\ 3\ 1\ 2\quad \times 4\ 2}$$
$$1\ 8\ 1\ 1\ 0\ 4$$

You have the final answer 181,104.

Example 3

Let us try another example. Multiply 9238 by 84.

$$9\ 2\ 3\ 8\quad \times 8\ 4$$

Step 1 - Add the same number of zeros as the number of digits in the multiplier.

$$\underline{0\ 0\ 9\ 2\ 3\ 8\quad \times 8\ 4}$$
$$\text{--}\ \text{--}\ \text{--}\ \text{--}\ \text{--}\ \text{--}$$

Step 2 - Sum of two pair products is a single digit of the answer.

Use your middle and forefinger to keep track of whether to calculate U or the T. Place each finger in the middle of two pairs.

Using 4 in 84, the pair product of 0 and 0 is 0 + 0 = 0 (U_4T_4 = 00 + 00).

Using 8 in 84, the pair product of 0 and 9 is 0 + 7 = 7 (U_8T_8 = 00 + 72).

Adding the two together we get 0 + 7 = 7.

$$U_4\ T_4$$
$$U_8\ T_8$$
$$0\ 0\ 9\ 2\ 3\ 8\quad \times 8\ 4$$
$$7\ \_\ \_\ \_\ \_\ \_$$

We move to the next two digits.

Using 4 in 84, the pair product of 0 and 9 is 0 + 3 = 3 (U_4T_4 = 00 + 36).

Using 8 in 84, the pair product of 9 and 2 is 2 + 1 = 3 (U_8T_8 = 72 + 16).

Adding the two together we get 3 + 3 = 6.

$$U_4\ T_4$$
$$U_8\ T_8$$
$$0\ 0\ 9\ 2\ 3\ 8 \quad \times 8\ 4$$
$$\overline{}$$
$$7\ 6\ \_\ \_\ \_\ \_$$

We move to the next two digits.

Using 4 in 84, the pair product of 9 and 2 is 6 + 0 = 6 (U_4T_4 = 3<u>6</u> + 0<u>8</u>).

Using 8 in 84, the pair product of 2 and 3 is 6 + 2 = 8 (U_8T_8 = 1<u>6</u> + <u>2</u>4).

Adding the two together we get 6 + 8 = 14.

$$U_4\ T_4$$
$$U_8\ T_8$$
$$0\ 0\ 9\ 2\ 3\ 8 \quad \times 8\ 4$$
$$\overline{}$$
$$7\ 6\ \_\ \_\ \_\ \_$$
$$+\ 1\ 4$$

Carrying over the 1, the 6 becomes 7.

$$0\ 0\ 9\ 2\ 3\ 8 \quad \times 8\ 4$$
$$\overline{}$$
$$7\ 7\ 4\ \_\ \_\ \_$$

We move to the next two digits.

Using 4 in 84, the pair product of 2 and 3 is 8 + 1 = 9 (U_4T_4 = 0<u>8</u> + <u>1</u>2).

Using 8 in 84, the pair product of 3 and 8 is 4 + 6 = 10 (U_8T_8 = 2<u>4</u> + <u>6</u>4).

Adding the two together we get 9 + 10 = 19.

$$\begin{array}{r} U_4\ T_4 \\ U_8\ T_8 \\ 0\ 0\ 9\ 2\ 3\ 8 \quad \times 8\ 4 \\ \hline 7\ 7\ 4\ \_\ \_\ \_ \\ +\ 1\ 9 \end{array}$$

Carrying over the 1, the 4 becomes 5.

$$\begin{array}{r} 0\ 0\ 9\ 2\ 3\ 8 \quad \times 8\ 4 \\ \hline 7\ 7\ 5\ 9\ \_\ \_ \end{array}$$

We move to the next two digits.

Using 4 in 84, the pair product of 3 and 8 is 2 + 3 = 5 (U_4T_4 = <u>1</u>2 + <u>3</u>2).

Using 8 in 84, the pair product of 8 and blank is 4 + 0 = 4 (U_8T_8 = 6<u>4</u> + --).

Adding the two together we get 5 + 4 = 9.

$$U_4\ T_4$$
$$U_8\ T_8$$
$$0\ 0\ 9\ 2\ 3\ 8 \quad \times 8\ 4$$
$$\overline{7\ 7\ 5\ 9\ 9\ \_}$$

We move to the next two digits.

Using 4 in 84, the pair product of 8 and blank is 2 + _ = 2 (U_4T_4 = 3$\underline{2}$ + --) and that becomes the last digit of the answer.

$$U_4\ T_4$$
$$U_8\ T_8$$
$$0\ 0\ 9\ 2\ 3\ 8 \quad \times 8\ 4$$
$$\overline{7\ 7\ 5\ 9\ 9\ 2}$$

Now you have the answer 775,992.

Example 4

Multiply 5743 by 63.

$$5\ 7\ 4\ 3 \quad \times 6\ 3$$

Take a second to apply the technique by yourself as fast as you can. Once you have the answer, you can check the steps below to see if you got your answer right.

Step 1 - Add the same number of zeros as the number of digits in the multiplier.

$$\underline{0\ 0\ 5\ 7\ 4\ 3 \quad \times 6\ 3}$$

$$\overline{-\ -\ -\ -\ -\ -}$$

Step 2 - Sum of two pair products is a single digit of the answer.

Use your middle and forefinger to keep track of whether to calculate U or the T. Place each finger in the middle of two pairs.

Using 3 in 63, the pair product of 0 and 0 is $0 + 0 = 0$ ($U_3T_3 = 0\underline{0} + \underline{0}0$).
Using 6 in 63, the pair product of 0 and 5 is $0 + 3 = 3$ ($U_6T_6 = 0\underline{0} + \underline{3}0$).

Adding the two together we get $0 + 3 = 3$.

$$U_3\ T_3$$
$$U_6\ T_6$$
$$\underline{0\ 0\ 5\ 7\ 4\ 3 \quad \times 6\ 3}$$
$$3\ \underline{\ \ \ }\ \underline{\ \ \ }\ \underline{\ \ \ }\ \underline{\ \ \ }\ \underline{\ \ \ }$$

We move to the next two digits.

Using 3 in 63, the pair product of 0 and 5 is $0 + 1 = 1$ ($U_3T_3 = 0\underline{0} + \underline{1}5$).

Using 6 in 63, the pair product of 5 and 7 is $0 + 4 = 4$ ($U_6T_6 = 3\underline{0} + \underline{4}2$).

Adding the two together we get $1 + 4 = 5$.

$$U_3\ T_3$$
$$U_6\ T_6$$
$$0\ 0\ 5\ 7\ 4\ 3 \quad \times 6\ 3$$
$$\overline{3\ 5\ \_\ \_\ \_\ \_}$$

We move to the next two digits.

Using 3 in 63, the pair product of 5 and 7 is 5 + 2 = 7 (U_3T_3 = 1<u>5 + 2</u>1).

Using 6 in 63, the pair product of 7 and 4 is 2 + 2 = 4 (U_6T_6 = 4<u>2 + 2</u>4).

Adding the two together we get 7 + 4 = 11.

$$U_3\ T_3$$
$$U_6\ T_6$$
$$0\ 0\ 5\ 7\ 4\ 3 \quad \times 6\ 3$$
$$\overline{3\ 5\ \_\ \_\ \_\ \_\ \_}$$
$$+1\ 1$$

Carrying over the 1, the 5 becomes 6.

$$0\ 0\ 5\ 7\ 4\ 3 \quad \times 6\ 3$$
$$\overline{3\ 6\ 1\ \_\ \_\ \_}$$

We move to the next two digits.

Using 3 in 63, the pair product of 7 and 4 is 1 + 1 = 2 (U_3T_3 = 2<u>1 + 1</u>2).

Using 6 in 63, the pair product of 4 and 3 is 4 + 1 = 5 (U_6T_6 = 24 + 18).

Adding the two together we get 2 + 5 = 7.

$$\begin{array}{r} U_3\,T_3 \\ U_6\,T_6 \\ 0\;0\;5\;7\;4\;3 \quad \times 6\;3 \\ \hline 3\;6\;1\;7\;\_\;\_ \end{array}$$

We move to the next two digits.

Using 3 in 63, the pair product of 4 and 3 is 2 + 0 = 2 (U_3T_3 = 12 + 09).
Using 6 in 63, the pair product of 3 and blank is 8 + _ = 8 (U_6T_6 = 18 + --).

Adding the two together we get 2 + 8 = 10.

$$\begin{array}{r} U_3\,T_3 \\ U_6\,T_6 \\ 0\;0\;5\;7\;4\;3 \quad \times 6\;3 \\ \hline 3\;6\;1\;7\;\_\;\_ \\ +1\;0\quad\; \end{array}$$

Carrying over the 1, the 7 becomes 8.

$$\begin{array}{r} 0\;0\;5\;7\;4\;3 \quad \times 6\;3 \\ \hline 3\;6\;1\;8\;0\;\_ \end{array}$$

We move to the next two digits.

Using 3 in 63, the pair product of 3 and blank is $9 + \_ = 9$ ($U_3T_3 = 0\underline{9} + --$) and that becomes the last digit of the answer.

$$U_3\ T_3$$
$$U_6\ T_6$$
$$\underline{0\ 0\ 5\ 7\ 4\ 3 \quad\quad x\ 6\ 3}$$
$$3\ 6\ 1\ 8\ 0\ 9$$

Now you have the answer 361,809.

If you got your answer wrong, don't worry. Just revisit the techniques and examples we covered in this chapter.

Read this chapter again if necessary. Then go to the practice section and complete the exercises. You might have understood the technique, but only practice will make using these techniques effortless and easy.

If you have any questions, you can ask it in the math Q & A section of the community by going **ofpad.com/mathqa** and we will make sure to respond. If you have not become a member of the community yet, watch the video **ofpad.com/communityguide** to get started.

If you have enjoyed the book so far, do leave a review on Amazon by visiting **ofpad.com/mathbook**.

If you did not enjoy the book so far, and if you have any general suggestions to improve the book or specific feedback for this chapter, do let us know at **ofpad.com/feedback**. If your feedback helps us improve the book even in small ways, we will thank you by sending a free review copy of our next product when it becomes available.

Once you finish practicing, move on to the next section.

Exercises

Download the rich PDFs for these exercises from **ofpad.com/mathexercises**.

01) 2951 x 7
02) 1315 x 7
03) 5127 x 6
04) 2934 x 9
05) 6999 x 8
06) 7326 x 7

07) 7469 x 96
08) 7770 x 73
09) 8626 x 69
10) 4025 x 70
11) 7731 x 63
12) 3196 x 63

13) 6671 x 86
14) 6402 x 29
15) 2185 x 93
16) 1919 x 83
17) 9372 x 64
18) 3374 x 99

Answers

01) 20,657
02) 9,205
03) 30,762
04) 26,406
05) 55,992
06) 51,282

07) 717,024
08) 567,210
09) 586,568
10) 281,750
11) 487,053
12) 201,348

13) 573,706
14) 185,658
15) 203,205
16) 159,277
17) 599,808
18) 334,026

Chapter 17 - LR Division

In this section we will look at how to do division using the LR Method.

Definition

To follow the steps in this section we need to understand some of the terminologies used in division.

- The number that is divided is called the dividend.
- The number which the dividend is being divided by is the divisor.
- The result we obtain is called the quotient.
- The number left over is called the remainder.

For example, if you divide 22 by 7 then 22 is the dividend and 7 is the divisor.

$$
\begin{array}{r}
3 \leftarrow \textbf{Quotient} \\
\textbf{Divisor} \rightarrow 7\overline{)22} \leftarrow \textbf{Dividend} \\
\underline{-21} \\
1 \leftarrow \textbf{Remainder}
\end{array}
$$

The result obtained which is 3 is the quotient and the number left over 1 is called the remainder.

You can also represent the same division in the following format where the remainder is represented as a decimal point.

$$\frac{22}{7} = 3.14...$$

Division Refresher

Before we look at how to do mental division, let us just brush up how we do division normally with a one-digit divisor. The following example is not any special mental math technique. You were probably taught to divide this way in school. We are covering this example as a refresher.

Let us say you want to divide 845223 by 4.

$$4 \overline{)845223}$$

We divide 8 by 4 to give quotient 2 with the remainder 0.

$$\begin{array}{r} 2 \\ 4 \overline{)845223} \\ \underline{-8} \\ 0 \end{array}$$

We divide 4 by 4 to give the quotient 1 with the remainder 0.

```
      21
    _____
4 ) 845223
    -8
    ──
    04
    -4
    ──
    0
```

We divide 5 by 4 to give the quotient 1 with the remainder 1.

```
      211
    _____
4 ) 845223
    -8
    ──
    04
    -4
    ──
    05
   -04
    ──
    1
```

We bring down the 2 and tag it along with the remainder 1.

```
      211
    _____
4 ) 845223
    -8
    ──
    04
    -4
    ──
    05
   -04
    ──
    12
```

We divide 12 by 4 to give the quotient 3 with the remainder 0.

```
         2113
    ┌─────────
  4 )  845223
       -8
       ──
        04
        -4
        ──
         05
        -04
        ───
          12
         -12
         ───
           0
```

We cannot divide 2 by 4 so we add a 0 to the quotient.

```
        21130
    ┌─────────
  4 )  845223
       -8
       ──
        04
        -4
        ──
         05
        -04
        ───
          12
         -12
         ───
           02
```

We divide 23 by 4 to get the quotient 5 and the final remainder of the division as 3.

```
        211305
    ┌─────────
  4 ) 845223
      -8
      ──
       04
       -4
       ──
        05
       -04
       ───
         12
        -12
        ───
          023
         -20
         ───
            3
```

You can stop here and write the answer of this division as 211305 with remainder 3 or you can write it as 211305 with remainder 3/4.

$$211305 \text{ r}3 \quad [OR] \quad 211305\frac{3}{4}$$

You can also continue to divide by adding a decimal point to the quotient.

We add a decimal point to the quotient and a 0 to 3, so it becomes 30.

```
        211305
      _____
   4) 845223
      -8
      ___
       04
       -4
       ___
        05
       -04
       ____
          12
         -12
         ___
          023
          -20
          ___
            30
```

30 divided by 4 gives us the quotient 7 with the remainder 2.

```
        211305.7
      _____
   4) 845223
      -8
      ___
       04
       -4
       ___
        05
       -04
       ____
          12
         -12
         ___
          023
          -20
          ___
            30
           -28
           ___
             2
```

We add a 0 to 2 so it becomes 20.

```
          211305.7
    4) 845223
       -8
       04
       -4
        05
       -04
         12
        -12
         023
         -20
          30
         -28
          20
```

Then we divide 20 by 4 to get the quotient 5 with the remainder 0.

We have the final answer in decimal point as 211305.75.

```
       211305.75
   4) 845223
      -8
       04
       -4
       05
      -04
        12
       -12
        023
        -20
         30
        -28
         20
        -20
          0
```

So this is how you will divide a number with a one-digit divisor.

When covering one-digit division in the future, we will not go over each individual step like how we did in this example and we will simply arrive at the quotient and the remainder.

For example, if we have to divide 845223 by 4 we will simply say the result is 211305.75 or we will say the quotient is 211305 and the remainder is 3 and not go over the individual steps.

$$\frac{845223}{4} = 211305.75$$

LR Division

There is no special trick required to do one-digit division because doing division with a one-digit divisor is simple if you know your multiplication table.

Also, you were probably taught to divide from left to right in school, so there is nothing new to learn there. So, we won't explicitly talk about the individual steps in the future examples.

Simply divide from left to right like how you were taught in school, when you have a one-digit divisor.

The remainder of this section will cover how to divide any number by a two-digit divisor which is a little trickier.

We will cover three different methods in this section and cover the more advanced flag pole method in the next section.

LR Division - Factoring

The first method we will cover is the division with factors.

It is possible to convert a two-digit or three-digit number into single digit factor and then do the division. It will greatly simplify the division process.

"Factors" are the numbers you multiply together to get another number.

Except for prime numbers, all other numbers can be broken down into smaller factors.

21 can be represented 7 x 3.

63 can be represented as 7 x 9 or 7 x 3 x 3.

Let us now see how we can simplify the division using factors.

To use factors to divide:

Step 1 - Break the divisor into smaller factors.

Step 2 - Divide the dividend by the smallest factor.

Step 3 - Then divide the resulting quotient with the larger factors to get your final answer.

Note: The factors should be lesser than or equal to 11 and you should know your multiplication table from 1 x 1 till 11 x 9 for you to be able to apply this technique.

Let us look at an example.

Example 1

Divide 512 by 16.

$$\frac{512}{16}$$

Step 1 - Break the divisor into one-digit factors.

So we break down 16 into its factors 4 x 4.

$$\frac{512}{4 \times 4}$$

Step 2 - Divide the dividend by the smallest factor.

Here both factors are 4 so it doesn't matter which factor you choose first to do the division.

512 divided by 4 gives us 128 as the first quotient.

$$\frac{512}{4} = 128$$

Step 3 - Divide the quotient with the larger factor to get your final answer.

So divide the quotient 128 by the second factor 4 to get 32 which is your final answer.

$$\frac{128}{4} = 32$$

Let us try another example.

Example 3

Divide 1376 by 32.

$$\frac{1376}{32}$$

Step 1 - Break the divisor into one-digit factors.

So we break down 32 into its factors 8 x 4.

$$\frac{1376}{8 \times 4}$$

Step 2 - Divide the dividend by the smallest factor.

Here 4 is the smaller factor. Dividing 1376 by 4 gives us 344 as the first quotient.

$$\frac{1376}{4} = 344$$

Step 3 - Then divide the resulting quotient with the larger factor to get your final answer.

So dividing the quotient 344 by 8 gives us 43 which is your final answer.

$$\frac{344}{8} = 43$$

In this example, you can simplify the division by 8 further by breaking down 8 into its factors 4 x 2. It will create an additional step but the division would be a lot easier to do. We will do that in the next example.

Example 3

Divide 3339 by 63.

$$\frac{3339}{63}$$

Step 1 - Break the divisor into one-digit factors.

So we break down 63 into its factors 7 x 9.

$$\frac{3339}{7 \times 9}$$

These are not the easiest factors to divide a number by. So let us break down 9 further into smaller factors 3 x 3.

$$\frac{3339}{7 \times 3 \times 3}$$

There are 3 factors (7 and two 3s) instead of two factors so we will now have to divide three times.

Step 2 - Divide the dividend by the smallest factor.

Here 3 is the smaller factor. Dividing 3339 by 3 gives us 1113 as the first quotient.

$$\frac{3339}{3} = 1113$$

Step 3 - Divide the resulting quotient with the next larger factor.

So dividing the quotient 1113 with the second 3 gives us 371.

$$\frac{1113}{3} = 371$$

We are not done because we have one more factor 7.

So dividing the quotient 371 by 7 gives us 53.

$$\frac{371}{7} = 53$$

All the examples we covered so far did not have decimal point. If the division results in a remainder, simply add a decimal point to the quotient and continue adding a zero to the remainder and keep calculating. Let us look at an example.

Example 4

Divide 546 by 16.

$$\frac{546}{16}$$

Step 1 - Break the divisor into one-digit factors.

So we break down 16 into its factors 4 x 4.

$$\frac{546}{4 \times 4}$$

Step 2 - Divide by the dividend by the smaller factor first.

Here 4 is the smaller factor. Dividing 546 by 4 gives us 136.5 as the first quotient.

$$\frac{546}{4} = 136.5$$

Step 3 - Then divide the resulting quotient with the larger factor to get your final answer.

So dividing the quotient 136.5 by 4 gives us 34.125 which is your final answer.

$$\frac{136.5}{4} = 34.125$$

Example 5

Divide 1381 by 32.

$$\frac{1381}{32}$$

Step 1 - Break the divisor into one-digit factors.

So we break down 32 into its factors 8 x 4.

$$\frac{1381}{8 \times 4}$$

Step 2 - Divide by the dividend by the smaller factor first.

Here 4 is the smaller factor. Dividing 1381 by 4 gives us 345.25 as the first quotient.

$$\frac{1381}{4} = 345.25$$

Step 3 - Then divide the resulting quotient with the larger factor to get your final answer.

So dividing the quotient 345.25 by 8 gives us 43.156. which is your final answer.

$$\frac{345.25}{8} = 43.156$$

Example 6

Divide 472 by 18. Calculate the quotient to two decimal places.

$$\frac{472}{18}$$

Take a second to apply the technique by yourself as fast as you can. Once you have the answer, you can check the steps below to see if you got your answer right.

Step 1 - Break the divisor into one-digit factors.

So we break down 18 into its factors 6 x 3.

$$\frac{472}{6 \times 3}$$

Step 2 - Divide by the dividend by the smaller factor first.

Here 3 is the smaller factor. Dividing 472 by 3 gives us 157.33 as the first quotient.

$$\frac{472}{3} = 157.33$$

Step 3 - Then divide the resulting quotient with the larger factor to get your final answer.

So dividing the quotient 157.33 by 6 gives us 26.22 which is your final answer.

$$\frac{157.33}{6} = 26.22$$

If you got your answer wrong, don't worry. Just revisit the techniques and examples we covered in this chapter.

Read this chapter again if necessary. Then go to the practice section and complete the exercises.

You might have understood the technique, but only practice will make using these techniques effortless and easy.

If you have any questions, you can ask it in the math Q & A section of the community by going **ofpad.com/mathqa** and we will make sure to respond. If you have not become a member of the community yet, watch the video **ofpad.com/communityguide** to get started.

Once you finish practicing, move on to the next section.

Exercises

Download the rich PDFs for these exercises from **ofpad.com/mathexercises**.

01) 6163 / 32
02) 824 / 63
03) 4162 / 36
04) 4597 / 24
05) 2775 / 12
06) 1524 / 10
07) 1777 / 45
08) 3301 / 8
09) 6984 / 72
10) 7865 / 9
11) 8936 / 12
12) 3771 / 72
13) 7591 / 54
14) 9599 / 72
15) 2185 / 42
16) 2461 / 63
17) 9691 / 18
18) 2561 / 42

Answers

01) 192.59
02) 13.08
03) 115.61
04) 191.54
05) 231.25
06) 152.4
07) 39.49
08) 412.63
09) 97
10) 873.89
11) 744.67
12) 52.38
13) 140.57
14) 133.32
15) 52.02
16) 39.06
17) 538.39
18) 60.98

Divisor Ends With 5

When the dividend ends with 5, it is easier to do the division if you double the dividend and the divisor.

When you do this, the divisor will become a single digit divisor.

Here are the steps:

Step 1 - If the divisor ends with 5, double the dividend & the divisor.

Step 2 - Do the division dropping the zeros in the end of the divisor.

Step 3 - Move the decimal to the left in the result by a number of places equal to the number of zeroes dropped in the divisor.

Let us look at an example.

Example 1

Divide 1195 by 35.

$$\frac{1195}{35}$$

Step 1 - If the divisor ends with 5, double the dividend & the divisor.

Since the divisor 35 ends with 5, we will double both the dividend and divisor.

Doubling the dividend 1195 we get 1195 x 2 = 2390

And doubling the divisor 35 we get 35 x 2 = 70

$$\frac{1195 \times 2}{35 \times 2} = \frac{2390}{70}$$

Step 2 - Do the division dropping the zeros in the end of the divisor.

Dropping the 0 in 70 we have 7.

2390 divided by 7 gives us 341.4

$$\frac{2390}{7} = 341.4$$

Step 3 - Move the decimal to the left in the result by a number of places equal to the number of zeroes dropped in the divisor.

Since we dropped one zero in 70, we will move the decimal place to the left by one-digit.

Moving the decimal point 341.4 becomes 34.14 and that is our final answer.

$$\frac{2390}{70} = 34.14$$

The original division was 1195 divided by 35 and 34.14 is the result of the division of the original problem.

$$\frac{1195}{35} = 34.14$$

Let us try another example.

Example 2

Divide 3423 by 75

$$\frac{3423}{75}$$

Step 1 - If the divisor ends with 5, double the dividend & the divisor.

Since the divisor 75 ends with 5, we will double both the dividend and divisor.

Doubling the dividend 3423 we get 3423 x 2 = 6846.

Doubling the divisor 75 we get 75 x 2 = 150.

$$\frac{3423 \times 2}{75 \times 2} = \frac{6846}{150}$$

Now we can proceed with the division by dropping the zero in 150 but resulting divisor 15 will still have two digits. Since the resulting 15 also ends with 5 we can double it again to get a single digit divisor.

So let us double both the dividend and the divisor again.

Doubling the dividend 6846 we get 6846 x 2 = 13,692.

And doubling the divisor 150 we get 150x 2 = 300.

$$\frac{6846 \times 2}{150 \times 2} = \frac{13692}{300}$$

Step 2 - Do the division dropping the zeros in the end of the divisor.

Dropping in the zeroes in 300 we have 3.

13,692 divided by 3 gives us 4564.

$$\frac{13692}{3} = 4564.0$$

Step 3 - Move the decimal to the left in the result by a number of places equal to the number of zeroes dropped in the divisor.

Since we dropped two zeros in 300, we will move the decimal place to the left by two digits.

So moving the decimal place by two places 4564 becomes 45.64 and that is our final answer.

$$\frac{13692}{300} = 45.64$$

45.64 is the result of the original division of 3423 by 75.

$$\frac{3423}{75} = 45.64$$

Let us try another example.

Example 3

Divide 3015 by 45.

$$\frac{3015}{45}$$

Take a second to apply the technique by yourself as fast as you can. Once you have the answer, you can check the steps below to see if you got your answer right.

Step 1 - If the divisor ends with 5, double the dividend & the divisor.

Since the divisor 45 ends with 5, we will double both the dividend and divisor.

Doubling the dividend 3015 we get 3015 x 2 = 6030.

And doubling the divisor 45 we get 45x 2 = 90.

$$\frac{3015 \times 2}{45 \times 2} = \frac{6030}{90}$$

Now proceed to do the division.

6030 divided by 90 is same as dividing 6030 by 9 and moving the decimal point to the left by a digit because there is one zero in 90.

6030 divided by 9 gives us 670. If you want to simplify the division, you can break 9 into its factors 3 x 3 and divide by 3 twice.

$$\frac{6030}{9} = 670$$

Since we are dividing by 90 and not 9, we will move the decimal place to the left by one-digit.

So moving the decimal place 670 becomes 67.0 and that is our final answer.

$$\frac{6030}{90} = 67.0$$

Our original problem was 3015 divided by 45 and this is the result of the division of the original numbers.

$$\frac{3015}{45} = 67.0$$

If you got your answer wrong, don't worry. Just revisit the techniques and examples we covered in this chapter.

Read this chapter again if necessary.

Then go to the practice section and complete the exercises.

You might have understood the technique, but only practice will make using these techniques effortless and easy.

If you have any questions, you can ask it in the math Q & A section of the community by going **ofpad.com/mathqa** and we will make sure to respond. If you have not become a member of the community yet, watch the video **ofpad.com/communityguide** to get started.

Once you finish practicing, move on to the next section.

Exercises

Download the rich PDFs for these exercises from **ofpad.com/mathexercises**.

01) 5363 / 25
02) 9353 / 25
03) 897 / 25
04) 8942 / 25
05) 7315 / 35
06) 2331 / 35

07) 5663 / 35
08) 4184 / 35
09) 5787 / 45
10) 5962 / 45
11) 1216 / 45
12) 2322 / 45

13) 7719 / 55
14) 4925 / 55
15) 7139 / 55
16) 5225 / 75
17) 7394 / 75
18) 6506 / 75

Answers

01) 214.52
02) 374.12
03) 35.88
04) 357.68
05) 209
06) 66.6

07) 161.8
08) 119.54
09) 128.6
10) 132.49
11) 27.02
12) 51.6

13) 140.35
14) 89.55
15) 129.8
16) 69.67
17) 98.59
18) 86.75

Factoring Fails with Prime Numbers

The techniques for division which we covered so far will work for most numbers. However, it might be difficult to apply these techniques for prime numbers greater than 11.

Prime numbers greater than 11 are the following: 13, 17, 19, 23, 29, 31, 37, 41, 43, 47, 53, 59, 61, 67, 71, 73, 79, 83, 89, and 97.

If the divisor is one of these numbers or a multiple of these numbers, then the methods we covered so far won't work.

For example, if the divisor is 31, 59 or 61, you will not be able to break it into smaller factors because these numbers are prime numbers.

Similarly if the divisor has one or more of its factors as prime numbers greater than 11, then the methods we have covered so far will not simplify the division.

For example, if the divisor is 34, 51 or 68, you will be able to break it into smaller factors like the following:

34 = 17 x 2

51 = 17 x 3

68 = 17 x 4

However, you will have not be able to simplify the division by 17 further because 17 is a prime number. So even though you will be able to factor the divisor, it won't really simplify the division.

If the divisor is 65, 85 or 95, then the trick of doubling the dividend and the divisor won't simplify the division because:

65 x 2 = 130

85 x 2 = 170

95 x 2 = 190

If you look closely 130, 170 and 190 are multiples of prime numbers 13, 17 and 19 that cannot be factored into smaller numbers. So, the method of doubling the dividend and the divisor won't simplify the division for all numbers.

It is difficult only because most of us don't commit the multiplication table of numbers greater than 11 to memory.

So what do we do in these cases?

LR Division With Rounding

When the divisor is a prime number or a multiple of a prime number, we might not be able to factor the prime number, but it is possible to round up or round down the divisor and then do the division. There is also the Flag Pole method which we will cover in the next section. So how do you apply this method?

To divide by rounding:

Step 1 - Round up or down the divisor if it is a prime number.

Step 2 – Divide the dividend with the rounded divisor. Don't calculate decimal places.

You can stop at **Step 2** if you only want an approximate answer.

Step 3 – Multiply the approximated quotient with the original divisor.

Step 4 – Subtract the product from the dividend & divide by the original quotient.

Step 5 – Add the result with the quotient to get your final answer.

Although you will be dividing by a two-digit number in **Step 4**, the result of the division will usually be a single digit number as long as you round-up or round-down to the nearest multiple of 10 in **Step 1**.

Let us look at a few examples to illustrate the method.

Example 1

Divide 2173 by 41.

$$\frac{2173}{41}$$

Step 1 - Round up or down the divisor if it is a prime number.

Here 41 is the divisor and it is a prime number. 41 can be rounded-down to 40.

$$\frac{2173}{40}$$

Step 2 – Divide the dividend with the rounded divisor. Don't calculate decimal places.

The rounded-down divisor is 40.

So we divide 2173 by 40 to get approximately 54. Note you don't have to calculate the decimal places.

54 is not the exact answer but it comes close, you can stop here if you only want an approximate result.

$$\frac{2173}{40} = 54$$

Step 3 – Multiply the approximated quotient with the original divisor.

If you multiplied 54 with 41 you would have got 2214.

$$\frac{2173}{41} \approx 54$$

$$54 \times 41 = 2214$$

Step 4 – Subtract the product from the dividend & divide by the original quotient.

Subtracting 2214 from 2173 we get 2173 - 2214 = -41.

$$2173 - 2214 = -41$$

Dividing -41 by the original quotient 41 we get -1. Although you are dividing by a two-digit number in this step, the result of the division will usually be a single digit number if you rounded-up or rounded-down the number to the nearest multiple of 10 in Step 1.

$$\frac{-41}{41} = -1$$

Step 5 – Add the result with the quotient to get your final answer.

Adding the result -1 with the quotient 54 we get 54 - 1 = 53 and that is our final answer.

$$\frac{2173}{41} = 53$$

Let us try another example.

Example 2

Divide 1827 by 29

$$\frac{1827}{29}$$

Step 1 - Round up or down the divisor if it is a prime number.

Here 29 is the divisor and it is a prime number. 29 can be rounded-up to 30.

$$\frac{1827}{30}$$

Step 2 – Divide the dividend with the rounded divisor. Don't calculate decimal places.

The rounded-up divisor is 30.

So we divide 1827 by 30 to get 60. Note you don't have to calculate the decimal places in this step.

60 is not the exact answer but it comes close. So, if you just want an approximate answer you can stop here.

$$\frac{1827}{30} = 60$$

Step 3 – Multiply the approximated quotient with the original divisor.

If you multiplied 60 with 29 you would have got 1740.

$$\frac{1827}{29} \approx 60$$

$$60 \times 29 = 1740$$

Step 4 – Subtract the product from the dividend & divide by the original quotient.

If you subtract 1827 from 1740 you will get 87. Subtract using the LR method.

$$1827 - 1740 = 87$$

If you divide the result 87 with the dividend 29 you get 3. Although you are dividing by a two-digit number in this step, the result of the division will usually be a single digit number.

$$\frac{87}{29} = 3$$

Step 5 – Add the result with the quotient to get your final answer.

Adding the result 3 with 60 we get 60 + 3 = 63 and that is our answer.

$$\frac{1827}{29} = 63$$

Let us try another example.

Example 3

Divide 3551 by 53.

$$\frac{3551}{53}$$

Take a second to apply the technique by yourself as fast as you can. Once you have the answer, you can check the steps below to see if you got your answer right.

Step 1 - Round up or down the divisor if it is a prime number.

Here 53 is the divisor and it is a prime number. 53 can be rounded-down to 50.

$$\frac{3551}{50}$$

Step 2 – Divide the dividend with the rounded divisor. Don't calculate decimal places.

The rounded-down divisor is 50.

So we divide 3551 by 50 to get 71. Note you don't have to calculate the decimal places in this step.

$$\frac{3551}{50} = 71$$

71 is not the exact answer but it comes close. So, if you just want an approximate answer you can stop here.

Step 3 – Multiply the approximated quotient with the original divisor.

If you multiplied 71 with 53 you would have got 3763.

$$\frac{3551}{53} \approx 71$$

$$71 \times 53 = 3763$$

Step 4 – Subtract the product from the dividend & divide by the original quotient.

If you subtract 3763 from 3551 you will get 3551 - 3763 = -212. Subtract using the LR method.

$$3551 - 3763 = -212$$

If you divide the result -212 with the dividend 53 you get -4. Although you are dividing by a two-digit number in this step, the result of the division will usually be a single digit number.

$$\frac{-212}{53} = -4$$

Step 5 – Add the result with the quotient to get your final answer.

Adding the result -4 from 71 we get 71 - 4 = 67 and that is our answer.

$$\frac{3551}{53} = 67$$

If you got your answer wrong, don't worry. Just revisit the techniques and examples we covered in this chapter.

Read this chapter again if necessary.

Then go to the practice section and complete the exercises.

You might have understood the technique, but only practice will make using these techniques effortless and easy.

If you have any questions, you can ask it in the math Q & A section of the community by going **ofpad.com/mathqa** and we will make sure to respond. If you have not become a member of the community yet, watch the video **ofpad.com/communityguide** to get started.

Once you finish practicing, move on to the next section.

Exercises

Download the rich PDFs for these exercises from **ofpad.com/mathexercises**.

01) 2436 / 13
02) 7004 / 17
03) 7803 / 19
04) 7004 / 23
05) 696 / 29
06) 839 / 31
07) 8532 / 37
08) 7984 / 41
09) 848 / 43
10) 9307 / 47
11) 1505 / 53
12) 3660 / 59
13) 7685 / 61
14) 8901 / 67
15) 968 / 71
16) 5384 / 79
17) 5908 / 83
18) 9631 / 97

Answers

01) 187.38
02) 412
03) 410.68
04) 304.52
05) 24
06) 27.06
07) 230.59
08) 194.73
09) 19.72
10) 198.02
11) 28.4
12) 62.03
13) 125.98
14) 132.85
15) 13.63
16) 68.15
17) 71.18
18) 99.29

Rounding With Reminder

The LR method with rounding will yield the exact result when there is no remainders or decimal places.

However, when the division results in a remainder you will have to estimate the decimal places of the answer.

The flag pole method which we will cover in the next chapter is more suited to get the exact answer along with the decimal places.

Let us look at one example on how to do this.

Example 1

Divide 3555 by 53.

$$\frac{3555}{53}$$

This division is almost same as the previous example but has a remainder of 4.

Step 1 - Round up or down the divisor if it is a prime number.

Here 53 is the divisor and it is a prime number. 53 can be rounded-down to 50.

$$\frac{3555}{50}$$

Step 2 – Divide the dividend with the rounded divisor. Don't calculate decimal places.

The rounded-down divisor is 50.

So we divide 3555 by 50 to get 71. Note you don't have to calculate the decimal places in this step.

71 is not the exact answer but it comes close. So, if you just want an approximate answer you can stop here.

$$\frac{3555}{50} = 71$$

Step 3 – Multiply the approximated quotient with the original divisor.

If you multiplied 71 with 53 you would have got 3763.

$$\frac{3555}{53} \approx 71$$

$$71 \times 53 = 3763$$

Step 4 – Subtract the product from the dividend & divide by the original quotient.

If you subtract 3763 from 3555 you will get 3555 - 3763 = -208. Subtract using the LR method.

$$3555 - 3763 = -208$$

If you divide the result -208 with the dividend 53 you get -3 with the remainder 50. Although you are dividing by a two-digit number in this step, the result of the division will usually be a single digit number, if you round up to the nearest multiple of 10 in Step 1.

$$\frac{-208}{53} = -3$$
Remainder: 50

Step 5 – Add the result with the quotient to get your final answer.

Adding the result -3 from 71 we get 71 - 3 which is equal to 68. Since there is a remainder we must also subtract the fraction 50/53 from 68. So the answer must lie between 67 and 68. Dividing 50/53 is almost 1. So subtracting 1 from 68 we will get approximately 67.

$$\frac{3555}{53} \approx 67$$

Use the LR Division with Rounding only when you want an approximate answer. Getting the exact answer when there is a decimal place is not always easy with the LR Division with

Rounding. The flag pole method which we will cover in the next chapter is more suited for that.

Combining Methods

When the divisor is a multiple of a prime number greater than 11, you combine the technique of factoring the divisor with rounding the divisor together.

Here are the steps:

Step 1 – Factor the divisor and divide the dividend using the smaller factor.

Step 2 – Divide the prime factor greater than 11, using LR Division with rounding.

We will look at just one example instead of three since this is a repetition of what we already covered.

Example 1

Divide 4346 by 82

$$\frac{4346}{82}$$

Factor the divisor and divide the dividend using the smaller factor.

Factoring 82 we get 2 x 41.

$$\frac{4346}{2 \times 41}$$

Dividing 4346 by 2 gives us 2173.

We are only left with 41 to complete the division and it is a prime number.

$$\frac{2173}{41}$$

Step 1 - Round up or down the divisor if it is a prime number.

Here 41 is a prime a number so 41 can be rounded-down to 40.

$$\frac{2173}{40}$$

Step 2 – Divide the dividend with the rounded divisor. Don't calculate decimal places.

The rounded-down divisor is 40.

So we divide 2173 by 40 to get approximately 54. Note you don't have to calculate the decimal places.

$$\frac{2173}{40} = 54$$

54 is not the exact answer but it comes close, you can stop here if you only want an approximate result.

Step 3 – Multiply the approximated quotient with the original divisor.

If you multiplied 54 with 41 you would have got 2214.

$$\frac{2173}{41} \approx 54$$

$$54 \times 41 = 2214$$

Step 4 – Subtract the product from the dividend & divide by the original quotient.

Subtracting 2214 from 2173 we get 2173 - 2214 = -41.

$$2173 - 2214 = -41$$

Dividing -41 by the original quotient 41 we get -1. Although you are dividing by a two-digit number in this step, the result of the division will usually be a single digit number.

$$\frac{-41}{41} = -1$$

Step 5 – Add the result with the quotient to get your final answer.

Adding the result -1 with the quotient 54 we get 54 - 1 = 53.

$$\frac{2173}{41} = 53$$

53 is the result of the division of 4346 divided by 82.

$$\frac{4346}{82} = 53$$

If you have any questions, you can ask it in the math Q & A section of the community by going **ofpad.com/mathqa** and we will make sure to respond. If you have not become a member of the community yet, watch the video **ofpad.com/communityguide** to get started.

If you have enjoyed the book so far, do leave a review on Amazon by visiting **ofpad.com/mathbook**.

If you did not enjoy the book so far, and if you have any general suggestions to improve the book or specific feedback for this chapter, do let us know at **ofpad.com/feedback**. If your feedback helps us improve the book even in small ways, we will thank you by sending a free review copy of our next product when it becomes available.

Chapter 18 - FP Division

In this chapter, we will look at how to do fast division using the flagpole method. This method might require you to write down the numbers, but with practice, you will be able to solve it without doing that.

Flag & Pole

Before we apply the flagpole method to division, it is important to understand which digit is the flag and which digit is the pole.

a) The first or the first set of numbers is the pole.
b) The second or the second set of numbers is the flag.

Let's look at a few numbers.

In 25, the pole is 2, and the flag is 5. It can be written like this 2^5.

In 83, the pole is 8, and the flag is 3. It can be written like this 8^3.

In 123, the pole is 12, and the flag is 3. It can be written like this 12^3.

Depending on how you group the number, in 123, you can also make 1 the pole and 23 as the flag. So it can also be written like this 1^{23}. You can also make 12 the pole and 3 the flag so it is written like this 12^3.

When To Use The FP Method

Just like the LR Division with Rounding, the flag pole method is useful when the divisor is a prime number greater than 11 or a multiple of a prime number greater than 11.

Prime numbers greater than 11 are the following: 13, 17, 19, 23, 29, 31, 37, 41, 43, 47, 53, 59, 61, 67, 71, 73, 79, 83, 89, and 97.

For divisors which are neither prime numbers greater than 11 nor multiples of prime numbers greater than 11, you can use LR Division with factoring.

Flag Pole method is an alternative to LR Division with rounding. With enough practice, you might find the flag pole method easier than LR Division with rounding.

Applying Flag Pole Method

To apply the flagpole method to division.

Step 1 - Divide the first digit of the dividend by the pole to find the first digit of the answer.

Step 2 - Attach the remainder from the previous step to the next digit of the dividend.

Step 3 - Multiply the last calculated digit of the answer with the flag and subtract it from the number arrived at after attaching the remainder.

Step 3b - If the number after subtraction is negative, reduce the last calculated digit of your answer by 1 and add the pole to the remainder before attaching it to the next digit of the dividend, and then repeat the previous step.

Step 4 - Divide the number from the previous step, by the pole to get the next digit of the answer.

Step 5 - If you cross the decimal line in the dividend add a decimal point.

Repeat steps 3 and 4 until all the digits of the answer are calculated.

Example 1

Let us now look at a few examples.

Divide 5578 by 31.

$$31 | \, 5 \; 5 \; 7 \, | \, 8$$

31 is a prime number so we will directly apply the flag pole method. In 31, we have 3 as the pole and 1 as the flag.

$$3^1 | \, 5 \; 5 \; 7 \, | \, 8$$

Step 1 - Divide the first digit(s) of the dividend by the pole to find the first digit of the answer

Divide 5 by 3 to get 1 with remainder 2

$3^1 | 5\ 5\ 7 | 8$
　　　$1\ \_\ \_\ \_$

Step 2 - Attach the remainder from previous step to the next digit of the dividend.

So attach 2 to the next digit.

$3^1 | 5\ {}_25\ 7 | 8$
　　　$1\ \_\ \_\ \_$

Step 3 - Multiply the last calculated digit of the answer with the flag and subtract it from the number arrived at after attaching the remainder.

So multiply the last calculated digit 1 with the flag 1 to get 1. Subtract it from 25 to get 24.

$$25 - 1 \times 1 = 24$$

Step 4 - Divide the number from previous step, by the pole to get the next digit of the answer.

We divide 24 by the pole 3 to get the next digit of the answer 8 and the remainder 0 is attached to the next digit.

$3^1 | 5\ {}_25\ {}_07 | 8$
　　　$1\ 8\ \_\ \_$

Step 3 - Multiply the last calculated digit of the answer with the flag and subtract it from the number arrived at after attaching the remainder.

So multiply the last calculated digit 8 with the flag 1 to get 8. If you try to subtract it from 7 you will get a negative number.

$$7 - 8 = -1$$

Step 3b - If the number after subtraction is negative, reduce the last calculated digit of your answer by 1 and add the pole to the remainder before attaching it to the next digit of the dividend, and then repeat the previous step.

So we reduce the last calculated digit 8 by 1 to get 7. We add the pole 3 to the remainder 0 to get 3. We then attach it to the next digit of the dividend 7 to get 37.

$$3^1 | 5 \ _{2}5 \ _{3}7 \ | \ 8$$
$$1 \ 7 \ \_ \ \_$$

Step 3 - Multiply the last calculated digit of the answer with the flag and subtract it from the number arrived at after attaching the remainder.

So multiply the last calculated digit 7 with the flag 1 to get 7. If you try to subtract it from 37 you will get 30.

$$37 - 7 = 30$$

Step 4 - Divide the number from previous step, by the pole to get the next digit of the answer.

We divide 30 by the pole 3 to get the next digit of the answer 10 and the remainder 0 is attached to the next digit.

Since the result of the division has two digits, we carry over the 1 so that 7 becomes 8 and we have the next digit of the answer as 0.

$$3^1 | 5 \ \ 2_5 \ \ 3_7 \ | \ 0_8$$
$$1 \ 8 \ 0 \ \_$$

Step 5 - Since we are crossing the decimal line, let us add a decimal point before carrying on.

$$3^1 | 5 \ \ 2_5 \ \ 3_7 \ | \ 0_8$$
$$1 \ 8 \ 0 \ . \ \_$$

Step 3 - Multiply the last calculated digit of the answer with the flag and subtract it from the number arrived at after attaching the remainder.

Even though, we carried over 1, the last calculated digits were 10. So multiply 10 with the flag 1 to get 10. Subtracting it from 8 gives us a negative number.

$$8 - 10 \times 1 = -2$$

Step 3b - If the number after subtraction is negative, reduce the last calculated digit of your answer by 1 and add the pole to the remainder before attaching it to the next digit of the dividend, and then repeat the previous step.

So we reduce the last calculated digit 0 by 1, so 80 becomes 79. We add the pole 3 to the remainder 0 to get 3. We then attach it to the next digit of the dividend 8 to get 38.

$$3^1 | \underline{5 \ 25 \ 37 \ | \ 38}$$
$$1 \ 7 \ 9 \ . \ \_$$

Step 3 - Multiply the last calculated digit of the answer with the flag and subtract it from the number arrived at after attaching the remainder.

So multiply the last calculated digit 9 with the flag 1 to get 9. If you try to subtract it from 38 you will get 29.

$$38 - 9 \times 1 = 29$$

Step 4 - Divide the number from previous step, by the pole to get the next digit of the answer.

We divide 29 by the pole 3 to get the next digit of the answer 9 and the remainder 2 is attached to the next digit.

$$3^1 | \underline{5 \ 25 \ 37 \ | \ 38 \ 2}$$
$$1 \ 7 \ 9 \ . \ 9$$

Since there are no more digits, we add an extra zero.

$$3^1 | \underline{5 \ 25 \ 37 \ | \ 38 \ 20}$$
$$1 \ 7 \ 9 \ . \ 9$$

Step 3 - Multiply the last calculated digit of the answer with the flag and subtract it from the number arrived at after attaching the remainder.

So multiply the last calculated digit 9 with the flag 1 to get 9. Subtract it from 20 to get 11.

$$20 - 9 \times 1 = 11$$

Step 4 - Divide the number from previous step, by the pole to get the next digit of the answer.

We divide 11 by the pole 3 to get 3 and 3 becomes the next digit of the answer.

$$3^1 | \; 5 \quad {}_2 5 \quad {}_3 7 \; | \; {}_3 8 \; {}_2 0$$
$$1 \; 7 \; 9 \, . \, 9 \; 3$$

If you want to calculate additional decimal points, then you can carry on this way by attaching the remainder to zero and repeating the steps.

Example 2

Let us now look at another example.

Divide 601324 by 73.

$$73 | \; 6 \; \; 0 \; \; 1 \; 3 \; 2 \; | \; 4$$
$$\underline{} \; \underline{} \; \underline{} \; \underline{}$$

73 is a prime number so we will directly apply the flag pole method. In 73, we have 7 as the pole and 3 as the flag.

$$7^3 | 6\ 0\ 1\ 3\ 2 | 4$$
$$\overline{}$$

Step 1 - Divide the first digit(s) of the dividend by the pole to find the first digit of the answer

We divide 60 by 7 to get 8 with remainder 4.

$$7^3 | 6\ 0\ _41\ 3\ 2 | 4$$
$$8\ \_\ \_\ \_$$

Step 2 - Attach the remainder from previous step to the next digit of the dividend.

So attach 4 to the next digit.

$$7^3 | 6\ 0\ _41\ 3\ 2 | 4$$
$$8\ \_\ \_\ \_$$

Step 3 - Multiply the last calculated digit of the answer with the flag and subtract it from the number arrived at after attaching the remainder.

So multiply 8 with the flag 3 to get 24. Subtract it from 41 to get 17.

$$41 - 8 \times 3 = 17$$

Step 4 - Divide the number from previous step, by the pole to get the next digit of the answer.

We divide 17 by the pole 7 to get the next digit of the answer 2 and the remainder 3 is attached to the next digit 3 of the dividend.

$$7^3 \overline{)6\ 0\ 41\ \ 33\ \ 2|\ 4}$$
$$8\ 2\ \_\ \_$$

Step 3 - Multiply the last calculated digit of the answer with the flag and subtract it from the number arrived at after attaching the remainder.

So multiply 2 with the flag 3 to get 6. Subtract it from 33 to get 27.

$$33 - 2 \times 3 = 27$$

Step 4 - Divide the number from previous step, by the pole to get the next digit of the answer.

We divide 27 by the pole 7 to get the next digit of the answer 3 and the remainder 6 is attached to the next digit 2.

$$7^3 \overline{)6\ 0\ 41\ \ 33\ \ 62|\ 4}$$
$$8\ 2\ 3\ \_$$

Step 3 - Multiply the last calculated digit of the answer with the flag and subtract it from the number arrived at after attaching the remainder.

So multiply the last calculated digit of the answer 3 with the flag 3 to get 9. Subtract it from 62 to get 53.

$$62 - 3 \times 3 = 53$$

Step 4 - Divide the number from previous step, by the pole to get the next digit of the answer.

We divide 53 by the pole 7 to get the next digit of the answer 7 and the remainder 4 is attached to the next digit 4.

$$7^3 | 6\ 0\ _41\ _33\ _62 | _44$$
$$8\ 2\ 3\ 7$$

Step 5 - Since we are crossing the decimal line, let us add a decimal point before carrying on.

$$7^3 | 6\ 0\ _41\ _33\ _62 | _44$$
$$8\ 2\ 3\ 7\ .$$

Step 3 - Multiply the last calculated digit of the answer with the flag and subtract it from the number arrived at after attaching the remainder.

So multiply the last calculated digit 7 with the flag 3 to get 21. Subtract it from 44 to get 23.

$$44 - 7 \times 3 = 23$$

Step 4 - Divide the number from previous step, by the pole to get the next digit of the answer.

We divide 23 by the pole 7 to get the next digit of the answer 3 and the remainder 2 is attached to the next digit.

$$7^3 | 6\ 0\ _41\ _33\ _62 | _44\ _2$$
$$8\ 2\ 3\ 7\ .\ 3$$

We add an extra zero since there are no more digits.

$$7^3 | 6\ 0\ 41\ 33\ 62 | 44\ 20$$
$$8\ 2\ 3\ 7\ .\ 3$$

Step 3 - Multiply the last calculated digit of the answer with the flag and subtract it from the number arrived at after attaching the remainder.

So multiply the last calculated digit 3 with the flag 3 to get 9. Subtract it from 20 to get 11.

$$20 - 3 \times 3 = 11$$

Step 4 - Divide the number from previous step, by the pole to get the next digit of the answer.

We divide 11 by the pole 7 to get the next digit of the answer 1.

$$7^3 | 6\ 0\ 41\ 33\ 62 | 44\ 20$$
$$8\ 2\ 3\ 7\ .\ 3\ 1$$

If you want to calculate additional decimal points, then you can carry on this way by attaching the remainder 4 to zero and repeating the steps.

Example 3

Let us now look at another example.

Divide 2829 by 123.

$$123 \overline{)\,2\ 8\ 2\ |\ 9\,}$$
$$\overline{}$$

123 is not a prime number. 123 can be factored as 41 x 3. You can divide the 2829 by 3 first and then use the flag pole method to divide the result by second factor 41.

$$\frac{2829}{123} = \frac{943 \times 3}{41 \times 3}$$

Take a second to apply the technique by yourself as fast as you can. Once you have the answer, you can check the steps below to see if you got your answer right.

First we divide 2829 by 3 to get 943.

$$\frac{2829}{3} = 943$$

If we divide 123 by 3 we get 41.

$$\frac{123}{3} = 41$$

So instead of dividing 2829 by 123 we divide 943 by 41.

So here 4 is the pole and 1 is the flag.

$$4^1 \overline{)\,9\ 4\ |\ 3\,}$$
$$\overline{}$$

Step 1 - Divide the first digit(s) of the dividend by the pole to find the first digit of the answer.

Divide the first digit 9 by the pole 4 to get 2 with the remainder 1.

$$4^1 | 9\ 4\ |\ 3$$
$$2\ \_\ \_$$

Step 2 - Attach the remainder from previous step to the next digit of the dividend.

Attach the remainder 1 with the next digit 4 to get 14.

$$4^1 | 9\ _14\ |\ 3$$
$$2\ \_\ \_$$

Step 3 - Multiply the last calculated digit of the answer with the flag and subtract it from the number arrived at after attaching the remainder.

Multiply the last calculated digit 2 with the flag 1 to get 2 and subtract it from 14 to get 12.

$$14 - 2 \times 1 = 12$$

Step 4 - Divide the number from previous step, by the pole to get the next digit of the answer.

Divide the number 12 by the pole 4 to get the next digit of the answer 3 with the remainder 0.

$$4^1 | \underline{9\ 14\ |\ 03}$$
$$2\ 3\ \_$$

Step 5 - If you cross the decimal line in the dividend add a decimal point.

Since we are crossing the decimal line, we will add a decimal point.

$$4^1 | \underline{9\ 14\ |\ 03}$$
$$2\ 3\ .\ \_$$

Step 3 - Multiply the last calculated digit of the answer with the flag and subtract it from the number arrived at after attaching the remainder.

Multiply the last calculated digit 3 with the flag 1 to get 3 and subtract it from 3 to get 0.

$$3 - 3 \times 1 = 0$$

Divide the number from previous step, by the pole to get the next digit of the answer.

Divide the number 0 by the pole 3 to get the next digit of the answer 0 with the remainder 0.

$$4^1 | \underline{9\ 14\ |\ 03}$$
$$2\ 3\ .\ 0$$

23 is the result of the dividing 2829 by 123.

If you got your answer wrong, don't worry. Just revisit the techniques and examples we covered in this chapter.

Read this chapter again if necessary. Then go to the practice section and complete the exercises.

You might have understood the technique, but it will take practice before the technique becomes second nature to you.

If you have any questions, you can ask it in the math Q & A section of the community by going **ofpad.com/mathqa** and we will make sure to respond. If you have not become a member of the community yet, watch the video **ofpad.com/communityguide** to get started.

If you have enjoyed the book so far, do leave a review on Amazon by visiting **ofpad.com/mathbook**.

If you did not enjoy the book so far, and if you have any general suggestions to improve the book or specific feedback for this chapter, do let us know at **ofpad.com/feedback**. If your feedback helps us improve the book even in small ways, we will thank you by sending a free review copy of our next product when it becomes available.

Once you finish practicing, move on to the next section.

Exercises

Download the rich PDFs for these exercises from **ofpad.com/mathexercises**.

01) 13|64924   04) 23|22046   07) 37|76989
02) 17|39466   05) 29|20046   08) 41|8692
03) 19|31993   06) 31|67401   09) 43|94187

10) 47|17080
11) 53|81357
12) 59|64453
13) 61|61927
14) 67|45042
15) 142|24546
16) 158|81448
17) 166|86864
18) 194|82474

Answers

01) 4994.15
02) 2321.53
03) 1683.84
04) 958.52
05) 691.24
06) 2174.23
07) 2080.78
08) 212
09) 2190.4
10) 363.4
11) 1535.04
12) 1092.42
13) 1015.2
14) 672.27
15) 172.86
16) 515.49
17) 523.28
18) 425.12

Chapter 19 - Eliminating Skill Atrophy

In this chapter, we will see how to eliminate skill atrophy. Any skill you learn will diminish over time if it is not used regularly. Regular practice is therefore vital. Let us look at the difference between students who did not practice and the students who practiced.

Students Who Did Not Practice

a) Never practiced after completing this book.
b) They remembered the techniques vaguely.
c) The speed with which they did mental math declined to the same speed as before they started the book.

Students Who Practiced

a) The students who practiced, on the other hand, practiced at least 10 minutes daily.
b) They completely absorbed the techniques.
c) The mental math skills became second nature to them.

Challenges You Will Face Practicing

There are two problems that come in the way of practising regularly.

1. First, you need the material to practice with.
2. Second, you need to remember to practice regularly.

There is a way to solve both these problems.

You can get practice workbooks delivered directly to your inbox every week. These workbooks will be delivered to your inbox for the next 8 weeks completely free.

Visit **ofpad.com/mathworkbook**. Just enter your email and click subscribe.

There are also alarm clock apps for Android and iOS that makes you solve math problems before you can turn them off. Unlike the weekly email workbooks, these apps won't let you practise every technique covered in this book. However, it will help you exercise some of your mental math skills everyday in the morning. I use the "I Can't Wake Up! Alarm Clock" for Android. There should be several other alternatives available.

Let us know in the math Q & A section how you plan to practice your new skills and master mental math **ofpad.com/mathqa**.

Chapter 20 - Conclusion

Congratulations on finishing this book and showing an initiative to learn and improve yourself.

Just keep practising, and you will be calculating lightning fast in no time.

If you have enjoyed the book, do leave a review on Amazon by visiting **ofpad.com/mathbook**.

If you are interested in receiving free review copies of our future products, let us know by going to **ofpad.com/reviewcopy**. When new books or courses come out, we will send you free copies for you to review.

If you did not enjoy the book, do give us some feedback at **ofpad.com/feedback** so we can improve. If your feedback helps us improve the book even in small ways, we will thank you by sending a free review copy of our next product when it becomes available.

If you haven't already done so, do check out the Mental Math Video Course available to enrol here **ofpad.com/mathcourse**. Someone who has benefited from this book will surely be able to get more out of the video course.

Since you purchased this book and showed an initiative to learn and improve yourself, the video course is available for you to purchase at $45 instead of the retail price of $295. If you are not happy with your investment in the Ofpad Video Course for any reason, there is a 30-day money back guarantee. So your investment in the video course will be 100% risk-free. Visit **ofpad.com/mathcourse** to enrol in the video course today.

Thank you for purchasing this book. I hope you have learnt something new and I wish you the best of luck in your mental math mastery. I will see you again as a reader/subscriber of Ofpad.com.

All the best.

About the Author

Abhishek V.R did his Mechanical Engineering in India and then went on to do his MSc degree in Business Analytics & Consulting from Warwick Business School in the UK. He now works in India. The author tries to continuously push the boundaries of his own intelligence. To do that, there are millions of books to read, thousands of podcasts to listen to and hundreds of seminars and workshops available to attend. They have the power to change lives. However, even with a burning passion to learn and absorb everything, most people lack the time, energy and resources to dedicate to constant learning and self-improvement. This has been the author's own personal struggle, and he hopes to provide a smart solution to this problem. Only 1% of the information available out there is required to make 99% of the difference in a person's life. The author creates content in his website Ofpad.com with the aim of sharing this 1% with the world. Since the author experiments with every dimension of his life which includes, health, productivity, habits, finances and fitness to just name a few, Ofpad should be of interest to anybody looking to improve their overall lives. You can learn more about the author and his work by visiting **Ofpad.com**.

Printed in Poland
by Amazon Fulfillment
Poland Sp. z o.o., Wrocław
20 December 2023

f719e5cd-df99-4f1d-b0b2-159168a1eec9R01